Apiculture in the Prehistoric Aegean

Apiculture in the Prehistoric Aegean

Minoan and Mycenaean Symbols Revisited

Haralampos V. Harissis
&
Anastasios V. Harissis

BAR International Series 1958
2009

Published in 2016 by
BAR Publishing, Oxford

BAR International Series 1958

Apiculture in the Prehistoric Aegean

ISBN 978 1 4073 0454 0

© The authors individually and the Publisher 2009

The authors' moral rights under the 1988 UK Copyright,
Designs and Patents Act are hereby expressly asserted.

All rights reserved. No part of this work may be copied, reproduced, stored,
sold, distributed, scanned, saved in any form of digital format or transmitted
in any form digitally, without the written permission of the Publisher.

BAR Publishing is the trading name of British Archaeological Reports (Oxford) Ltd.
British Archaeological Reports was first incorporated in 1974 to publish the BAR
Series, International and British. In 1992 Hadrian Books Ltd became part of the BAR
group. This volume was originally published by John and Erica Hedges Ltd. in
conjunction with British Archaeological Reports (Oxford) Ltd / Hadrian Books Ltd, the
Series principal publisher, in 2009. This present volume is published by BAR
Publishing, 2016.

Printed in England

BAR titles are available from:

           BAR Publishing
           122 Banbury Rd, Oxford, OX2 7BP, UK
EMAIL   info@barpublishing.com
PHONE +44 (0)1865 310431
FAX   +44 (0)1865 316916
           www.barpublishing.com

For Pascale and Sofia
H.H

Table of contents

Preface v
Introduction 1
Part 1: The literary sources and the archaeological evidence 3
Apiculture in prehistory 3
Apiculture in prehistoric Aegean 9
Beekeepers and beekeeping practices 13
Beekeeping paraphernalia 18
Beehives 18
Smoking pots 27
Vessels from the "Snake Room" in Knossos 29
Part 2: The iconographical evidence. Minoan and Mycenaean symbols revisited . 39
Contra Evans 39
"Fruits on sacred trees" 43
"Festoons and pearls" 46
"Ritual bending of the sacred tree" 48
"Flying gods" 50
"Great Goddess" 51
"Baetyls and bird epiphany of deities" 53
"Altars" 57
"Horns of concentration" 62
"Sacral knot" 66
"Temenos" 66
"Double axe" 68
Epilogue 71
Appendix 73
Works Cited 83

Preface

This work is based on several years of personal research but its completion would not have been possible without the help of friends along the way. We wish to express our warmest thanks and respects to Kostas Zachos, Angelika Douzougli, Sarah Morris and John Papadopoulos for their encouragement and support in the research and writing of this monograph. We also thank Kostas Tatsis and Theodoros Karalis for their advice on the subject of modern and traditional beekeeping. The results of this monograph do not necessarily express the opinion of the aforementioned. We are most grateful to Professor Ingo Pini for supplying drawings of the rings and seals and to Vassiliki Gregoropoulou, Eustratia Roidou and Chariklia Naka for bibliographical assistance. We thank Vasilios A. Harissis for drawings, assistance and together with Sofia H. Devletoglou-Harissi for lifelong guidance.

Abbreviations for primary texts, authors and periodicals follow S. Hornblower and Antony Spawforth, eds., *Oxford Classical Dictionary*, Third Edition Revised, Oxford University Press, 2003.

Permissions and courtesies of the figures were obtained as follows: Figures 1, 5, 8(3), 11, 12, 13(4), 13(6), 27, 28, 29, 36(3,4,5), 46(3), 51(3),57(2) with permission by Gerald Duckworth & Co. Ltd); Figures 2, 23, 30, 41, 53, 59(1,2,4,5,6) courtesy Herakleion Museum, Ministry of Culture, Archaeological Receipts Fund; Figures 4, 8(1), 8(2), 10(1), 38(4), 48(2,3), 51(1,2), 52 courtesy T. Bikos; Figures 20, 21, 22, 58 courtesy V. Harissis; Figure 7(1) courtesy of Robert Laffineur; Figure 7(2) courtesy American School of Classical Studies at Athens; Figure 15 courtesy of Archaeological Society at Athens; Figures 31(1,2,3,4,7,8,9,10,11,12,13,14,15,16,17,18,19,20,21,22,23,24) courtesy Corpus der Minoischen und Mykenischen Siegel (CMS), Marburg; Figures 31(5,6) courtesy Ashmolean Museum/Ms. Helen Hughes Brock; Figure 55 courtesy P. Kamilakis.

Introduction

Ever since antiquity, people highly valued bee and the hive products; the bee for its industrious and organizational capabilities, honey for its nutritious and therapeutic qualities and beeswax and propolis for their multiple uses. Early historic texts reference bees in order to clarify particular aspects of human life, such as the relationship between man and woman (Hes. *Theog.* 594-602), as exemplars of a community (Cic. *Off.* 1.157) and as models for the desired social order (Xen. *Oec.* 7.32). Pliny (*HN* 11.11) reserved his unbounded admiration for them, even ascribing to them a *res publica, consilia* and *mores* and Aristotle (*Gen. an.* 761a, 5) believed that bees are of holy nature. Hilda Ransome in *The Sacred Bee in ancient times and folklore* reviewed how this unceasing admiration has been expressed through the ages. Malcom Fraser in *Beekeeping in Antiquity* reviewed and magnificently commented the ancient literary sources concerning beekeeping activities in Greece and Rome. Eva Crane in *The world history of Beekeeping and honey hunting* has thoroughly and meticulously collected the archaeological evidence concerning beekeeping paraphernalia. She dedicated a chapter in Greece concerning findings dated in classical and later periods.

The few things known about prehistoric apiculture come mainly from Egyptian depictions and they have all been collected and presented by Crane. What is evident from these depictions is the remarkable preservation of apiculture techniques and paraphernalia from prehistoric up to modern times. However, for prehistoric Aegean no such a systematic review exists, despite several isolated but important findings of beekeeping paraphernalia in prehistoric stratums. Having studied these findings and by comparing them with modern folkloric apiculture items we have ascertained that as in Egypt so too in Greece there was but very little change of the forms of beekeeping items since prehistoric times. We subsequently expanded our research to other folkloric apiculture paraphernalia and also in methods, habits and beliefs still in use by the traditional beekeepers in Greece and Eastern Mediterranean in order to investigate for as yet unrecognized apiculture prehistoric Aegean items and methods. A valuable guidance to our research was the above mentioned work of Eva Crane and the work (in papers and congresses) by Thanasis Bikos, who has thoroughly studied traditional beekeeping in Greece.

Our research has been surprisingly fruitful and in the following we present our results which however intriguing might appear are nevertheless the result of a coherent and consistent method that we have carefully followed throughout our work.

Part 1: The literary sources and the archaeological evidence

Apiculture in prehistory

Honey is a sweet and viscous fluid produced by bees from the nectar of flowers[1] and laid down in their nests to be used as their source of energy when food sources are scarce. Man always considered honey of great value as a sweetener, preservative for foods and an ingredient of multiple therapeutic remedies[2] (Porph. *De antr. nyph.* 15.6). Honey was also supposed to provide longevity (Ath. 3.59; Democritus ap. *Geoponica* 15.7) and its high nutritional value rendered it a worthy *"food of the gods"* (Varro *Rust.* 3.16.5)[3]. For Hesiod (*Op.* 232–4) bees are ranged alongside sheep as the most important features of the hillsides, reinforcing the view that humanity would not exist without honey[4]. The underpinnings of such beliefs are the proven therapeutic qualities of honey, as well as the fact that honey yields, with relatively little work, an alimentary guarantee, even in arid places such as the islands of the Aegean rich as they were with thyme and other important plants utilised by bees, e.g. phrygana[5].

For all these qualities honey has been highly praised since prehistory. As early as the sixth dynasty an Egyptian official named Sebni takes honey on his expedition to

---

[1] Nectar is a sugar-rich liquid produced by the flowers of plants in order to attract pollinating insects. It is produced in glands called nectaries, which are generally at the base of the perianth, so that insects-pollinators are forced to brush the flower's reproductive structures, the anthers and pistil, while accessing the nectar. Nectar is the sugar source for honey

[2] Clark 1942. For the uses of honey in Egypt from the period of Old Kingdom, see Lucas 1999, 25-26. In ancient Greece, honey constituted the basic component of hundreds of pharmaceutical recipes. For the medicinal uses of honey in ancient Greece and the Byzantine period, see Varella 2000, 221-231; Crane 2000a, 514. For the therapeutic uses of honey in the Islamic world see Fahd 1968, 73-4. For honey as a food preservative in the Islamic world see Fahd 1968, 74, n 65. In the Roman period although sugar was not unknown it was only used for therapeutic purposes (Plin. *HN* 12.17) (see Lucas 1999, 25). Sugar cane, a plant from India, reached Mesopotamia in the 6th century. From there it was made known to all the Moslem world and Syria, Morocco, Spain, Sicily, and Egypt. From Egypt it was later introduced to Greece and Europe (Lombard 1994, 33-34, 40, 180-181, 185, 212). Much later, sugar cane was imported to Crete by Venetians (Kaplan 1992, 26 n 11,12)

[3] *Ambrosia* and *nectar*, by which the gods were nourished, were probably honey and mead (a drink containing honey) (Weniger 1884, 2638; Fraser 1951, 125-28). For the nutritional value of honey see Dumay 1997, 75-78

[4] Dumay 1997, 76, 78

[5] Petanidou 1995; Tselios 1997. For the distribution of phrygana in Greece, see Polunin 1987, 36-7. Virgil (*G* 4.127) speaks of a piece of land that was inadequate for tillage or for planting of vines but which an apiarist managed to render quite suitable for his bees. Plato (*Critias* 111c) reports some regions *"which have nothing but food for bees"*. A Cypriot dictum values honey for its high nutritional qualities and the relatively little outlay that necessitates (Rizopoulou-Igoumenidou 2000, 407).

exchange it for other goods. In the *Papyrus Harris* king Ramses III enumerates his gifts to the important temples which include thirty thousand jars of honey and three thousand deben of beeswax. Certain officials of the Middle Kingdom received monthly gifts of royal honey and they proudly mentioned this in their inscriptions[6].

In Anatolia too, honey was highly valued and produced in great quantities; at around 2000 BCE, Sinuhe reports a district of Palestine that *"copious was its honey"*. Honey figures largely amongst the booty collected by the Egyptian kings during their campaigns in Palestine and Syria. From his fifth campaign Thotmes III brings back tribute of 470 jars of honey, during his seventh campaign he collects *"regular supplies of honey"* and from his fourteenth campaign 246 jars of honey were captured in Syria[7]. Jacob had been able to trade honey for the corn of Egypt (*Genesis* 43.11) and in later days Ezekiel reminds Tyre that *"Judah and the land of Israel were thy merchants: they traded in thy market [...] and honey, and oil, and balm"* (*Ezekiel* 27.17). Honey until recent times was considered as a food fit only for kings and lords[8].

Other bee products such as wax and propolis ("bee-glue") had multiple uses in prehistory[9]. Beeswax was used for providing light[10] and as a coating for wooden plates used for engraving texts (*POxy* 736.16; Lib. *Ep.* 886.1)[11]. It was also used in painting (encaustic technique) (*IG*42(1).102.272 - 4th c. BCE) , the manufacture of boats, the production of adhesive substances[12] and as a cosmetic (Philostr. *Ep.* 22). Propolis was used as a component of cosmetics[13] and, together with wax, it formed an adhesive substance[14].

Hence, apiculture in antiquity was a most important and profitable activity highly valued by both elites and non-elites. In prehistoric Egypt, it was an exclusive privilege of

---

[6] During the reign of Thotmes II, Ineni was *"supplied from the table of the king, with honey"* (Forbes 1966, 81-82)

[7] Forbes 1966, 81-82

[8] Honey is mentioned as the food of sovereigns in Greek traditional Christmas carols (*"honey for the sovereigns and milk for the lords"*) and in Greek traditional riddles (Hatzaki-Kapsomenou 2001, 312). Great quantities of honey were reserved for king's use in Sudan and Ethiopia through the 18th and 19th centuries (Newberry 1938). The *Edict of Rothair* in the Lombardic Laws imposed a fine of twelve solidi for stealing another man's hive. The importance of the offence may be judged from the fact that no other kind of theft was punished more severely in the Edict (*Edict of Rothair*, sections 318-9 in Drew 1973, 114). In Anglo-Saxon England the fines for bee-theft were quite serious. Alfred reformed the laws, making fines for theft consistent, but he explicitly says that bee-theft was originally a crime with greater penalty than others, only comparable to the theft of gold (Libermann 1903, 54-5)

[9] Propolis is a wax-like resinous substance collected by honeybees from tree buds or other botanical sources and used as cement to reinforce the structural stability of the hive and to seal cracks in the hive. "Typical" propolis has approximately 50 constituents, primarily resins and vegetable balsams (50%), waxes (30%), essential oils (10%), and pollen (5%). For prehistoric use of wax and propolis see Crane 2000a, 524, 527; Lucas 1999, 89, 337

[10] Evidence for wax use in Egypt is provided by tomb paintings dating to the New Kingdom (1567-1085 BCE) (Crane 2000a, 524, fig. 49.2a). On the use of candles in Roman times and later, see Crane 2000a, 525, 600

[11] Such as the one found on the Ulu Burun shipwreck (Bass 1990)

[12] Lucas 1999, 2-3

[13] Lucas 1999, 89

[14] Lucas 1999, 3

the Pharaoh to have beehives in his property[15]. Pharaohs of Lower Egypt from the 4th Dynasty through the 4th c. BCE bore the title of "apiarist"[16] and the bee hieroglyph denoted the Pharaoh of Lower Egypt from the 1st Dynasty up to the Roman period (Horapollo *Hieroglyphica* 1.62; Ammianus Marcellinus *Historiae* 17.4.11; Chaeremon[17]).

Pharaohs assigned the control of beekeeping products to high ranking officials. During the 1st Dynasty such an official bore the title "*Sealer of Honey*"[18] while titles as "*Nomarch, Royal Acquaintance, and Overseer of Beekeepers*" and "*Overseer of Beekeepers of the Entire Land*" are also known from the Middle Kingdom[19]. An official responsible for beekeeping is also known from the New Kingdom; he was called Rekhmire and was a vizier of Upper Egypt in the reign of Tuthmosis III[20]. In his tomb, frescoes were found with apiculture scenes. One of his duties was to receive offerings (cereals, honey and all other "valuable products") at the temple on behalf of the Treasury, and to stamp them according to their quality. Such a seal (bearing four bees) for marking vessels with honey is known from this period[21].

We learn for another duty of such officials from a 19th-dynasty papyrus from Luxor, now in the Ashmolean Museum: a scribe complained of two beekeepers who had made inadequate returns with respect to their honey production. One of them still continued working although he had been dismissed, and the scribe urged his superior (obviously an overseer of beekeepers) to take firmer action[22].

Fig.1. Mesolithic rock painting at Ganeshghati, near Bhopal, central India showing Apis dorsata nests in a tree (Register NA-02, Sharma & Ali, 1980 from Crane 2000a, Fig. 10.2a)

Bee and honey also played an important part of the temple ritual, notably that of the god Min. The priest helping the high-priest of Coptos to prepare the unguent for the god Min was called the "*beekeeper*" and an official of the reign of Ramses IV calls himself "*purveyor of the honey of Min*"[23]. By the Ptolemaic period, royal but private as well bee farms are known to have existed[24].

---

[15] Beinlich 1994; Crane 2000a, 170; Kueny 1950, 85 and 93. Beehives were also the property of temples as the temple of Isis (*Pap. Petrie* 3.43)

[16] Chouliara-Raios 1989, 32-34; Gardiner 1957, 71-8; Ransome 1937, 26

[17] Chaeremon is reported by Leclant 1968, 56. See also Ransome 1937, 26; Shaw 2003, 6-7

[18] Ransome 1937, 26

[19] Martin 1971, 234

[20] Davies 1944

[21] Now in the Agricultural Museum in Dokki (Crane 2000a, 165). In Mesopotamia bees are depicted on seals from Uruk III (ca. 3000 BCE) (Theodorides 1968, 5)

[22] Crane 2000a, 165

[23] Montet 1950. Also known are the "*beekeepers of Amon*" (Lefebvre 1929, 46). In *Arabic News* (www.arabicnews.com) (07/10/2002) is reported "A Supreme Council of Antiquities (SCA) mission unearthed a Middle Kingdom (15 century B.C.) tomb that belongs to Epi, the beekeepers supervisor in god Amon's house, said SCA Secretary General Zahi Hawas".

[24] Martin 1971, 234

Honey was initially collected from wild bee nests in tree and rock cavities or from tree branches where the wild Mediterranean-Levantine bees (*Apis mellifera*) and the bees of Anatolia and India (*Apis florae* and *Apis dorsata*) would, respectively, make their hives. This practice is represented in rock paintings of the Paleolithic period from Spain and elsewhere and of the Mesolithic period from Spain and India[25] (fig.1). Bee domestication, i.e. organised apiculture with beehives, was invented later, when the quantity of honey and other wild bee products did not suffice for the large population settlements of the Neolithic period. It is possible that during the Neolithic, the control of production passed to local chieftains. The Egyptians were the earliest culture known to keep bees in hives. This activity is first attested in a relief of the Fifth Dynasty (although the origin of this occupation must reach much farther back into antiquity[26]) at the "Chamber of the Seasons" in the solar temple of Niuserre at Abu Ghurab. The relief shows a man kneeling in front of a pile of cylindrical hives, exhaling into it. This has been interpreted as "blowing" smoke into the hive as modern Egyptian beekeepers do to draw out the bees and access the honey[27]. Evidence from the Veda implies that developed apiculture existed also in prehistoric India[28], and in Hittite tablets of 1500 BCE there are legislative regulations concerning apiculture with beehives[29].

Against such a backdrop, the existence of organised (with beehives) apiculture in the prehistoric Aegean is not unreasonable. Honey, wax and mead ($\mu\epsilon\lambda\acute{\iota}\tau\epsilon\iota\sigma\varsigma$ $\sigma\acute{\iota}\nu\sigma\varsigma$), are mentioned in Linear B tablets and probably in Linear A as well[30]. Archaeological evidence confirms that in Late Helladic, honey, wax and mead were in use[31]. Written records imply that Crete was, as it still is, a place famous for its honey a fact that is easily explained by its abundance in thyme; Nicander (ap. Columella *Rust.* 9.2) states that apiculture was discovered in Crete during the time of Saturn. The drowning of mythical Glaukos (son of Minos) in a big jar of honey indicates significant honey production while the fable of

---

[25] Crane 2000a, 36-40

[26] Crane 2000a, 170-1. The first written record for honey dates to the Old Kingdom (Lucas 1999, 25, n 8)

[27] Keuny 1950; Crane 2000a, 161-171; Jones et al. 1973, 403. Beehives are represented also on the following: a relief found during the excavations of the Causeway of Unas (c. 2350 BCE); a wall painting in Tomb 100 of Rekhmire (West Bank, Luxor) (c. 1450 BCE); a wall painting in Tomb 73 from the same region; and an incised and painted relief in Tomb 279 of Pabesa (West Bank, Luxor) (c. 664-625 BCE) (Crane 2000a, 163-67, fig. 20.3a, 20.3b, 20.3c, 20.4a, tb. 20.3A; Kueny 1950)

[28] Dave 1954/55

[29] Goetze 1955, 193

[30] Honey: PY *Un* 718.5; KN *Gg* 702; 704; 36 (Olivier 1973, 122). The ideograms *132 and *211 correspond to amphorae of honey (Davaras 1984, 77). Beeswax: Mc 142; Ma* 44; KN U 172; U 436; U 746 (Chadwick 1973, 50-51, Evershed et al. 1997). Mead: PY Wr 1360. cf Melas 1999, 489-490. Linear A: Sign L 64 (Myres I. L. *Scripta Minoa* II, 1952, 19 AB 55; Grumach 1964, 7-14; Hood 1976, 67f n.48, fig 1,15)

[31] For evidence of wax see Evershed et al. 1997. Glotz (1923, 183) considers likely the use of candles in Minoan Crete because of the discovery of "candlesticks" in Knossos (see Evans 1921-1935, I, 578-9, fig. 422, 423a, 423b). Glotz (1923, 169) also believes that in Minoan Crete existed apiarists with the emblem of a gloved hand and a bee and that these apiarists transmitted their knowledge to mainland Greece. For evidence of mead use see Tzedakis 1999: 133 (No 115), 167 (No 153, 154, 155), 169 (No 158, 159), 170 (No 161), 176 (No 167, 168, 169)

Daidalos, who built wings using beeswax (Ov. *Met.* 8.183), presupposes considerable quantities of wax. The myth of the bronze giant Talos, may refer to the creation of a statue by Hephaistos (Simon. 568 PMG) or Daidalos (Soph. fr. 160, 161 R) using the lost-wax casting[32] which would also require large quantities of wax. The testimonies that the Telchines of Crete made the first-ever statues (Zeno fr. 1 ap. Diod. Sic. 5.55.1-3; Pind. *Ol.* 7.50-53; Callimachus *Hymn* 4, 30-31; Nicolaus fr. t116) is probably also connected to this technique. Archaeological records show that lost-wax casting was already in use in Judea in 4th millennium, and it is suggested that it was used in Mesopotamia, in Lemnos (Poliochni - Early Bronze Age IIb)[33] and in Crete[34]. Crete, in Roman times, was famous for the large quantities of honey and the quality of wax it produced (Plin. *HN* 11.14, 11.21). In Byzantium honey from Sammonion promontory in Eastern Crete[35] was considered one of the best in the Mediterranean (*Geoponica* 15.7). During the Arab occupation Cretan honey continued to be produced in large quantities and was exported to Egypt[36]. The first written record of beehives in the Greek world, however, concerns Archaic Athens, where Solon issued laws specifying the minimum distance between the beehives of neighbouring apiarists (Plut. *Sol.* 23.8)[37].

Most of the extant apiculture texts date to the Roman period, in authors like Varro, Virgil, Columella, Palladius, and Hyginus, but there are numerous references made by these writers to others who wrote apicultural treatises before them, such as Democritus, Menecrates of Ephesus, Aristotle, Mago of Carthage and Nicander. Some authors, such as Aristomachos from Soli in Cyprus and Philiskos of Thasos, wrote treatises having studied apiculture for many decades (Plin. *HN* 11.9), thus proving the existence of advanced knowledge on the subject much earlier than the Roman period. Aristotle's *Historia animalium* includes several brilliant pages of detailed bee biology knowledge. It is even speculated, that Aristotle (*Hist. An.* 624b) knew of the so-called "dance of bees" much earlier than Karl von Frisch's Nobel Prize winning description[38]. Was this advanced Greco-

---

[32] As suggested by Cook 1964, 723-4. For the description of the technique see Lucas 1999, 221

[33] See for references Crane 2000a, 530 and DeJesus 1980, 126

[34] Faure 1999, 373

[35] Probably to be identified with modern Cape Sideron in Eastern Crete (Sanders 1982, 1/38, 1/40)

[36] Genesios, ed. Lesmuller-Werner, Thurn, 33.3 in Christides 1984, 98, 117

[37] For the Roman *horoi* that defined the limits of an apiary on Mt. Hymettos in Attika see Ober 1981 (but cf Langdon 1985)

[38] Haldane 1955. The common opinion today is that the ancients did not know the sex of the queen which was first mentioned by Queen Elizabeth's I's beekeeper Charles Butler in his *The Feminine Monarchie* (1609) and by the Dutch physiologist Ian Swammerdam (1637-1680) whose observations on a microscope were published after his death in a Latin translation (*Biblia Naturae* 1737). Actually, in 1634, before the discovery of Swammerdam, Thomas Mouffet (*Theatrum Insectorum*. London) knew of the copulation of queen with the drones. But the so-called *Pascale candle* quotes a passage from the late 3d century in which St. Ambrose invokes the *"blessed and marvellous mother bee"* (Crane 2000a, 591). Furthermore in "swarm charms" of the 9th c. from Western Europe the sovereign of bees is called "*mother*" (Ransome 1937, 165 and Crane 2000a, 591, Spamer 1978, Fife 1964). Swarm charms were short verses used by beekeepers to persuade an airborne swarm to settle. They were widely used by traditional beekeepers in Greece and one always finds in them the calling of the

Roman knowledge a late import from elsewhere or the product of a long accumulation of information from locally practiced apiculture?

---

bee-mother (Giousaris 2000, 316, Kostakis 1963, 388). In ancient literature, the kings of bees (ἡγεμόνες) usually characterize the drones. Xenophon (*Oec.* 7.17) however, reports η εν τω σμήνει ἡγεμών μέλιττα and Arrian (*Historia Indica* 8.2) speaks of βασιλέα ἢ βασίλισσαν of bees and elsewhere (*Epict. Diss* 3.22.99) of the "queen" of bees. Aristotle (*Hist. an.* 553a) reports that the sovereigns of bees "are called mothers because they give birth" (καλοῦνται ὑπό τινων μητέρες ὡς γεννῶντες). For wasps also he says that the ἡγεμόνες are called μήτρες (mothers) (*Hist. an* 628a). In the work Ἰωσήφ καὶ Ἀσενέθ of 2nd c. μέλισσαι μεγάλαι ὡς βασίλισσαι ("*large bees such as the queens*") are reported showing the knowledge that the queen who is bigger than the workers was known as such and she was called actually queen and not king. From the above we may reach the conclusion that the ancients, or at least some of them, knew of the existence of a queen/mother bee. See also Hudson - Williams 1935

Apiculture in prehistoric Aegean

It seems that bees had a special importance for the Minoans and Mycenaeans, having been found on jewels discovered in Mycenaean graves (Mycenae grave III, Late Helladic III Peristeria grave II, Late Helladic III Thebes Shaft grave)[39] while the oldest and most impressive image of a bee on jewellery is probably the Middle Minoan I pendant from Chrysolakkos grave of Mallia[40] (fig. 2).

The Mallia pendant that resembles the Egyptian jewel of Senusret II (1897-1878 BCE)[41], depicts two bee-like insects holding in their third pair of legs a ball while in their mouth, as well as between their antennae, they bear spherules. Additional spherules are hanging from their wings and their bodies. Opinions on whether the pendant depicts bees or wasps are divided[42]. Kitchell[43] has shown that the pendant recalls the report by Aristotle (*Hist. An.* 554a.17, 624a) that the bee carries the "wax" and the pollen with its legs and vomits the honey from its mouth in a comb cell.

Fig. 2. Gold pendant from Mallia (photo Herakleion Museum)

---

[39] Peristeria grave II: Marinatos S. *ΠΑΕ* 1962, 97, tb 99a; *ΑΔ* 18, 1963, B, 100, tb 121; *SMEA* 3, 1967, 12, fig. 20. Thebes Shaft grave: Buchholz 1973, 184, no 11

[40] Kitchell 1981. Davaras (1984, 82) dates it in 1700 BCE

[41] Hood 1976, 72

[42] Supporters of bees: Tzavella-Evjen 1970, 127 who supports that the pendant portrays the copulation of a king bee, Buchholz 1973, 183f; Richards-Mantzoulinou 1979, 7; Kitchell 1981, 9. Davaras (1984, 83) also reports Schering W. 1976. *Funde auf Kreta*, 108-10; Effenterre van H.1980. *Le palais de Mallia et la cite minoenne*, 498. Supporters of wasps: Higgins R.A. 1962, *JHS* 82:198; Catling H. 1963. *JHS* 83:214; Warren P. 1975.*The Aegean Civilizations* 41; Hiller S. 1981/2. *AfO*, 28: 191; Platon N. 1954. *ΠΑΕ*, 364-5; Lafleur R.A., Matthews R.W., McCorkle D.B. 1979. "A Re-examination of the Mallia Insect Pendant" *AJA* 83: 208-12

[43] Kitchell 1981 where also some other improbable hypothesis (honeycomb, wood pulp, dung ball)

This is an almost perfectly accurate observation since it is known that the foraging worker bees transport the liquid nectar in their stomach and once inside the beehive, they pass it from mouth to mouth to other receiving worker bees which process it further and store it eventually in a comb cell as honey[44].

Pollen (gr. ἐριθάκη; lat. *pollen, polis, pulvis*, in the sense of fine flour, sometimes called flower sperm or simply "flower") is a fine to coarse powder consisting of microgametophytes, which carry the male gametes of seed plants[45]. Pollen, rolled into a ball, is stored and transferred in pouches (pollen baskets) on the hind legs (third pair) of the bee just as can be shown in the Mallia pendant and as is recorded by Pliny (*HN* 11.10): "*The bees whose business it is to carry the flowers, with their fore feet load their thighs, which Nature has made rough for the purpose*". But unlike pollen, wax is secreted by glands on the ventral surface of the abdomen of the young worker bees – it is not carried on the legs as, besides Aristotle, Aristophanes too wrongly thinks in *Wasps* (107-8):"*to* [...] *return home with his nails full of wax like a bumble-bee*". Aristophanes' and Aristotle's "wax" should refer to propolis, a wax-like resinous substance, which is collected from tree buds or other botanical sources[46] and is carried back to the beehive in special receptors on the hind legs of the bee, just like pollen[47] (fig. 3). We believe that it is Pliny's (*HN* 11.10) phrase that completes the description of the pendant: "[the bees] *then return to the hive, where there are three or four bees ready to receive them and aid in discharging their burdens*". When the foraging worker bees reach the hive other worker bees unload the propolis from them, but not the pollen which the worker bees dispose directly in the cells without help from other bees[48]. The bees in the hive

Fig. 3*. Pollen carrying bee*

---

[44] Winston 1987, 99

[45] Pollen is produced in the anther. The transfer of pollen grains to the female reproductive structure (pistil) is called pollination. This transfer can be mediated by the wind, in which case the plant is described as anemophilous (literally wind-loving) or by insects and in this case the plant is described as entomophilous (literally insect-loving). Entomophilous plants produce pollen that is relatively heavy, sticky and protein-rich, for dispersal by insect pollinators attracted to their flowers. Many insects and some mites are specialized to feed on pollen, and are called palynivores. Honey mixed with pollen are the main diet of young bees (called larvae)

[46] Propolis was called by the ancients *"tear drops"* of the trees and flowers (Aristotle *Hist. an.* 553b.28; 623b 29, Virgil *G*. 4.160, *Geop*. 15.3.3). Dioscorides (*De materia medica* 2.84.1) calls it πρόπολις and Hesychius (s.v.) σαμυλίς. Aristotle also calls it μίτυς and πισσόκηρος (*Hist. an.* 624a). Πισσόκηρος means wax from pitch (πίσσα) the resin-drop of conifer trees that Hesychius (s.v) calls *"tear of pines"*. Also called κηρόπισσος by Hippocrates (*De Morbis i-iii* 2.18) and κήρωσις by Aristotle (*Hist. An*. 553b). Virgil (*G*. 4.160) calls it *gluten* and Pliny (*HN* 11.14) *melligo*

[47] Winston 1987, 23-28, fig. 3.11, fig. 3.12

[48] Winston 1987, 24

unload the in-coming foraging bees by using their mandible not their feet as one can see in the pendant. But such divergence is easily explained by the fact that the pendant is a piece of art never intended to serve as a scientific rendering of the bees; depicting in this way the bees, the symmetry of the pendant is better served. The "droplets" on the wings of the bees, like the one between the antennae, if not simply pollen balls adherent on their bodies[49], could represent drops of water since the ancients believed that bees carried water in their mouths and on their bodies: "[the bees] *charge the downy surface of their bodies with drops of liquid*" (Plin. *HN* 11.10). Actually the foraging worker bees carry the water in their stomach and when they reach the beehive they pass it with their mouth to the mouth of other workers[50].

Thus, we believe that the Mallia pendant depicts two worker-bees, heavily charged with pollen (or propolis)[51] in their hint legs, nectar[52] (or water) in their mouth and pollen (or drops of water) on their bodies, helping each other as described by Aristotle and Pliny[53]. It is reasonable to assume that it was not the wasp but the hard working bee - an object of human fascination and delight for eons - that inspired the admiration of the prehistoric artist of the Mallia pendant just as was the case for numerous poets, philosophers and theologians.

Such detailed knowledge, as evidenced by the Mallia pendant, argues for active apiculture in the Bronze Age. Renfrew[54] hypothesised the existence of organised apiculture in the Aegean on the basis of the indirect testimony of a Protocycladic terra-cotta bear that appears to be feeding on honey from a vessel (a beehive?), and some EHII/III seals from Lerna depicting a bee. According to Renfrew these seals were used for sealing containers of honey.

Bees have also been depicted in a number of Minoan rings, seals and sealings: the Koumassa ivory ring CMSII1,159 (EMII-MMIA), the Krassi three-sided steatite prism CMSII2,225, the Metropolitan four-sided chalcedony prism CMSXII,109 (MMIII) and the three-sided CMSXII,117, the black marble three-sided prism bead CMSXII,42 (MMI), the steatite seal CMS XII,142 (MMIII), and CMS VII,15b, the black jasper lentoid CMS VII,70; the seal from Mallia/Mu HM 2390, the three-sided prism from Ajios Onouphrios CMSII1,111, the three-sided prism form Sammlung Metaxas CMSIV,22D, the Candia three-sided prism P.1, the Siteia four-sided bead-seal P.29, the British Museum lentoid of

---

[49]Traditional apiarists in Cappadocia used to say that the foraging bees have "their heads and their legs full of flowers" meaning the pollen that adheres on the body of the bee (Kostakis 1963, 386). Dots adhering to the body of a bee are portrayed, according to Evans (1909, 167, fig. 86b), in a bee ideogram from Crete

[50]Winston 1987, 121

[51] The granular aspect of the ball makes pollen ball more likely than propolis

[52] Kitchell (1981) identifies the ball in the mouth of bees as honey

[53]Some authors (as Kitchell 1981, 11) have proposed that the pendant shows Cretan bees that carry a pebble, which serves as a stabiliser against being drifted by the wind, as described by Plutarch (*De soll. An.* 967b) and others (Plin. *HN* 11.10; Arist. *Hist. an.* 626b 24; Verg. *G.* 4.194; Ael. *NA* 1.11, 5.13; Ambr. *Virginit.* 106). Although it is a precise observation it is an erroneous interpretation because the particular kind of bee that uses these small pebbles, the mason bee, uses them to build its hive and not as a stabilizer (Allaby 1998 s.v "mason bee")

[54] Renfrew 1972, 287

black jasper CMSVII,70 (LM), the Kasarma lentoid CMSV,579 (LM), CMS VII,71, CMS VII,86, CMS IV,79 (MMI); the Phaistos sealings CMSII5,312, CMSII5,313, CMS II5,314, CMS II5,315 and CMS II5,316 (MMIIB), the sealing from Mallia/Mu" HM 1090, and the inscriptions on clay from Malia/Mu HM 1664 (MMII) and from Mallia/P H 18 and H e[+]f (MMIII)[55]. Symbol 34 "🐝" on the MMIIB Phaistos Disk is also considered to represent a bee[56]. The hieroglyphic sign 86 (with characteristic resemblance with those of Egypt[57]) occurring on a three-sided prism seal from the Mirabello province (P.20 = CMS IX,21D), in a sealing from the hieroglyphic deposit of the Palace at Knossos (CMSII8,62) (MMII), in four inscribed legends on clay labels also from the aforementioned hieroglyphic deposit (P.51a = CMSII8,80, P.54b, P.76a, P.86b) (MMII) led Evans to believe that a widespread beekeeping industry existed in Minoan Crete[58]. An impression with a bee from a triangular seal from Karahoyuk near Konya dating c. 1750 BCE must also be assigned to Cretan influences[59].

In Middle Minoan Phaistos, Mallia and Knossos, existed a fully-developed system incorporating sealings into a bureaucratic organization, an offshoot, and a very close one, of the widespread and ancient sealing system which was found throughout the Near East[60]. Rings and seals seem to belong to officials responsible for the security of objects or products of certain value. Jars, sacks, matting, storeroom doors, chests and whatever contained goods were sealed with clay nodules and the same applied for documents. Although the relationship between a seal's iconography and its owner's administrative function has not been much studied, some scholars assume a direct relationship[61]. Hence, in the case of seals bearing a bee sign it would not be unreasonable to suppose that the sealed product was honey and/or wax and the owners of the rings were Minoan officials with a role similar to the Egyptian official called *Sealer of Honey* or *Overseer of Beekeepers*. Grave offerings too, as the Mallia pendant, were probably chosen among the personal belongings of the deceased because they were recognized as typical signs of his/her belonging to a specific class, or signs of their exercise of specific functions a custom known from other contemporary Minoan burials[62].

---

[55] CMS VII,86 is recognized as "a bee" in Kenna 1967; CMS XII,142 is recognized as "a bee" in Kenna 1972; CMS IV,79 is recognized as "a bee" in Sakellarakis 1969; CMS II5,314 and CMS II5,316 are recognized as "a bee" in Pini 1970
[56] Duhoux 1977, reported by Woudhuizen 1997, 100
[57] Woudhuizen 1997
[58] Evans 1921-1935, I, 281
[59] Alp 1968, 175, no 50, Abb. 66
[60] Weingarten 1990, 105
[61] Kilian-Dirlmeier 1987; Wingerath 1995, 147-8
[62] For a bronzesmith, a carpenter, a leatherworker, a lapidary and a weaver Minoan burials signalled by grave goods see Renfrew 1972, 341-342; see also Laffineur 1992, 110

Beekeepers and beekeeping practices

The title of the apiarist (*me-ri-te-wo*) appears among the Linear B Pylos tablets[63] as a person of some importance, perhaps actually an official in charge of honey production[64].

According to later literary sources both men and women engaged in apiculture. All of them were always of a noble ancestry and of a high social status recalling the Egyptian status of the prehistoric beekeepers overseers: Erechtheus (or Erichthonios as he is called alternatively but actually the same person according to Σ*Hom. Il.* 2.547) who came from Egypt (Diod. Sic. 1.29.1) in 16th c. BCE (according to the *Marmor Parium*) was a king who introduced apiculture in Attica (Columella *Rust.* 9.2); Boutes, high priest and brother of Erechtheus, bears a name that resembles the Egyptian word "bjtj"[65], a title of an official, a priest, responsible for "honey-hunting"[66]. Both of them were descendants of king Kekrops who also came from Egypt (Phylochorus fr. 93; Theopompus ap. Eusebius *Evangelic Preparation* 10.10.23; Charax ap. Σ*Lycophron* 111) and he too, introduced apiculture in Attica (Verg. *G.* 4.177). Another prominent male figure in beekeeping mythology (dated by ancient sources in 16th c. BCE[67]), was Aristaios, son of Apollo. He introduced apiculture in mainland Greece and in Kea from Libya. Aristaios was taught the art of apiculture by women: the nymphs Vrisai or Vlisai (Arist. fr. 511; Diod. Sic. 4.81.2; Oppianus *Cynegetica* 4.271)[68]. Other nymphs, the daughters of Phryxon, reared the bees that nourished Jupiter in the Dictaen Cave in Crete (Euhemerus ap. Columella *Rust.* 9.2.3)[69]. The nymphs were considered the sisters or the female counterparts of Kouretes (Hesiod *Cat.* fr. 10a. 17-19), themselves often called Kourai. Cretan Kouretes [(Κούρητες or Κωρῆται[70] or Κώρειτες (Corinna fr. 1a)] were, among other things, well known beekeepers, as Diodorus (5.65) and Justinus (*Epit.* 44.4.1) record. They appear already in Homer and Hesiod and they prevail until late antiquity but only in Ephesus and in Crete[71]; in the rest of the Greece they constituted a memory of a remote "mythical" past.

Their groups consisted of both men and women (Diod. Sic. 5.66). Strabo (10.3.11) and Hesiod (*Cat.* fr. 10a.17-19) identify or closely relate the Kouretes with the Satires who according to Ovid (*Fast.* 3.735) had discovered apiculture. Satyrs were associated with Pan who was the protector of apiculture (*Anth. Pal.* 16.189, Theophylactus *Epistulae* 2; Sardianos *Anth. Pal.* 9.226, 6-7; Theoc. *Id.* 5.59) and in classical antiquity, ceramic

---

[63] PY *Ea* 481; 771; 799; 801; 813; 820 (Olivier 1973)
[64] Chadwick 1976, 124-6
[65] Richards-Mantzoulinou 1979, 85, n 52
[66] Montet 1950, 25; *Lexikon der Agyptologie* s.v *biene*
[67] According to ancient testimonies (Hes. *Th.* 977), Aristaios was a contemporary of Kadmos, who, according to the *Marmor Parium* (*IG* 12(5), 444), lived in 1519 BCE (LHI). Janko (1982, 248 n 38) and Triomphe (1989, 35-38) believe that the fable of Aristaios refers to prehistoric times and is older of the colonisation of Cyrenaica by settlers from Thera but cf West (1985, 85-87). For more on Aristaios see Appendix
[68] Many aspects in the story of Aristaios can be recognized as apiculture practices that still prevail among traditional beekeepers (see Appendix)
[69] For more "myths" that make bees the nymphs' proxies see Larson 1995
[70] *IC* I:xxv.3, *IC* I:xxxi.7 (1st c. BCE) (Crete)
[71] IC I.xxv.3; IC I.xxxi.7 and 8 (in Willetts 1962, 209)

beehives were being placed in many caverns of Attica where Pan's cult was usually practiced[72].

Some "rituals" of the Kouretes were in fact purely apiarian practices such as the characteristic "fable" which attests them striking their bronze shields during the hour that Jupiter was born in a cave, where he was nourished by the bees in order "to cover the baby's cries from hostile Saturn" (Epimenides fr. 22; Callimachus. *Hymn* 1, 49-50; Korinna 654i.12-18 PMG; Verg. *G.* 4.149; Diod. Sic. 5.70.5; *Myth. Vat.* 2. 3). This particular "ritual" of the Kouretes has been identified with the practice of apiarists to attract swarms by striking bronze objects (tanging)[73]. Ovid (*Fast.* 3.735) reports that Satyrs attracted swarms by tanging. This practice was known to Aristotle (*Hist. An.* 627a.15) and it is also reported by Roman (Plin. *HN* 11.22; Didym. *Geop.* 15.3.7; Ael. *NA* 5.13; Varro *Rust.* 3.16.7; Vg. *G.* 4.64; Columella *Rust.* 9.12) and Byzantine sources (*Geoponica* 15.3.7). In the Suda this subterfuge is called μελιττοπηχεῖν and is still mentioned in modern European books of apiculture and practiced worldwide[74].

Capturing a swarm during the swarming period [which Columella (*Rust.* 9.14.5) and Palladius (*Opus agriculturae* 7.7.4) rightly place between the rising of the Pleiades (in the end of May according to the Julian calendar) and the summer solstice (end of June)] is of vital importance to apiarists in maintaining their bees since the loss of a swarm amounts to a significant loss of honey and beeswax. Hence the English dictum "*a swarm of bees in May, is worth a cow and calf that day*"[75]. Hittites' laws record that a swarm of bees was worth the same as a sheep[76]. Consequently apiarists during the swarming period must be vigilant about recovering the swarm by directing it to an empty hive or capturing it from the tree. Ancients had various ways of directing a swarm to an empty beehive or even preventing it from leaving the hive. Aristotle (*Hist. An.* 627b) warned that when the bees were hanging as clusters at the entrance of the beehive they were about to leave. He cautioned apiarists to blow sweet wine on the beehive to avert the departure of the swarm. The same practice was followed by the traditional apiarists in Greece, who kept in their mouth the Sacral Communion (which is made of sweet wine) and blew it on the hives[77]. This habit was so widespread during the Byzantine period that forced the Orthodox Church to publish a decree prohibiting it[78]. The efficacy of this practice can be explained by the fact that bees are using sugar as food and since sugar in the form of sweet wine is provided to them, the swarm would not leave since the main reason for swarming is shortage of food

---

[72] Wickens 1986, 194

[73] Willetts 1962, 217 n 122; Willetts 1985

[74] Cotton 1842, 338 (reported by Ransome 1937, 226); Fraser 1951, 46. This practice exists also in Pontos (Topalidis 1968/1969, 336), in Germany (Ransome 1937, 175) but also among the Bushmen of Africa (Bleek 1911, 353 reported by Ransome 1937, 298). In Cappadocia (Kostakis 1963, 388) they used to strike pieces of a broken pottery a practice also reported by Aristotle (*Hist. an.* 627a 15). *Geoponica* 15.3 report also the "*rhythmical clamping of the hands*"

[75] Reported by Ransome 1937, 230. See also the corresponding French dictum "*un essaim du mois May, vaut une vache du pays Bray*"

[76] Goetze 1955, 193; Crane 2000a, 173

[77] Kukules 1951, 356; Vrontis 1939, 203, 208

[78] Kukules 1951, 356

or of space in the hive. Other means for not loosing a swarm can not be explained biologically. They concern the so-called bee charms as the throwing of earth in the air to make the swarm settle (Verg. *G.* 4.87; Varro *Rust.* 3.16.30; Plin. *HN* 30.53). This practice was widespread in Greece, in Cyprus[79] and in other parts of Europe[80] until recently. The use of musical sounds to attract swarms was also common among beekeepers[81]. Whistling to the bees is attested in the Old Testament (*Isa* 7.18) and in modern Greece (Naxos)[82]. In 19th c. England it was said that if one sang to his bees they would not leave the apiary[83]. In Chalkidike they used to call the swarms loudly[84]. Ancient Egyptian beekeepers "called" the bees with a reed flute[85]. We think that the tanging performed by the Kouretes should also be ranked in the category of these musical bee-charms.

One of the Kouretes was called "*Melisseas*" (Apollod. *Bibl.* 1.5; Zenobius 2.48; Nonnus Dion. 13.145, 28.306, 30.305, 32,271, 36.280, 37.494 .520 .534 .675 .702; Diod. Sic. 5.61) a name that characterizes today the beekeeper in Chalkidike ("Meliss[e]as")[86] and in Linear B, as we saw above, *me-ri-te-wo* is a title of the official in charge of honey production. Personal names originating from old professions have survived in Byzantium[87] and in modern Europe and Greece[88]. We know from the Linear B tablets that the same thing was also happening in prehistoric Greece: dozens of personal names in the Mycenaean vocabulary represent armorers, carpenters, weavers, dyers, unguent boilers and other artisans[89]. Similarly, in Babylonia, Assyria and in Ancient Palestine, persons' names were often designated by their profession[90]. Vrisai, the women that taught the art of apiculture to Aristaios had probably a similar status with Kouretes Melisseas since their name can be etymologised from the Cretan word $βριτύ$ meaning sweet (Hsch. s.v.) or the verb $βλίττειν$ meaning honey harvesting (Pl. *Rep.* 8.564e)[91].

But this apiculture could refer to a primitive type of honey collection: the collection of honey from the trees, which, as we mentioned, existed since the Palaeolithic period. Honey-hunters existed in Egypt since at least the time of Ramses IV (12th c. BCE)[92]. The practice existed also in historical times, as testified by Aesop (*Fabulae Aphthonii rhetoris* 2 and *Fabulae dodecasyllabi* 85) and Claudian (*In Ruf.* 2.460), by Pseudo-Aristotle (*Mir. ausc.* 831b.26) for Lydia and by Diodorus (5.14.1) for Corsica. Virgil (*G.* 2.452) and the Suda (s.v. $ἀκρίς$) report also a primitive tree-beekeeping which

---

[79] Vrontis 1939, 203, Panaretos 1967, 281
[80] Betts 1922; Spamer 1978; Fraser 1951, 46
[81] For medieval and later bee-charms (musical and non-musical) used by apiarists in Europe and in Greece see Fife 1964; Spamer 1978
[82] Oikonomidis 1965-1966, 632
[83] Cowan 1865, 188-90
[84] Petropoulos 1957, 191
[85] The habit it is attested at least from the 5th c. BCE (Leyden Papyrus *Myth of the eye of Ra* Cat. 1384, cited by Crane 2000a, 170)
[86] Petropoulos 1957, 192
[87] Kukules 1951, 352
[88] Loukatos 1992, 196
[89] Morpurgo Davies 1979
[90] Mendelsohn 1940, 18
[91] Chantraine s.v.
[92] *Papyrus Harris* 28.3, 46.1, 48.2 in Breasted 1962, part 4 § 266; Montet 1950, 25

according to Mnaseas (fr. 5) was taught to humans by bee-priestess, the "*priestess Melissa*". The tree that entertains most often the hives of wild bees is oak since apart from the fact that its flowers are honey-producing it has also cavities in its trunk (Aesop *Fabulae Aphthonii rhetoris* 27; Theophr. fr. 190). According to Nicander (*Alex.* 445), the first hive of bees was in an oak tree. The Roman poets speak of the honey flowing from oaks during the "Golden Period" of Saturn (Verg. *Ecl.* 4.30, *G.* 1.131; Tib. *Elegiae* 1.3.45; Ov. *Am.* 3.8.40, *Met.* 1.112; Triphiodorus 534) and Hesiod (*Op.* 233) "*οὔρεσι δε δρῦς ἄκρη μέν τε φέρει βαλάνους, μέσση δέ μελίσσας*" refers to wild bees nests in oak cavities. One Hittite text may be referring to this fact when it describes the arboreal residents of a tree: "*Above an eagle perched on* (its) *branches, below a snake coiled about its trunk* (?), *in* (its) *midst a bee hove*" (KUB 43.62 iii 57'). The scholia in Nicander (ΣNicander 448c 2) and Pseudo-Phocylides (*Opinions* 173), both report as an old habit of apiarists the collection of honey from oak cavities. On a 6th c. BCE black amphora of unknown provenance now in Antikenmuseum of Basel, honey hunters are represented being attacked by wild bees during their attempt to collect honey from a tree. In Roman period, the existence of such honey-hunters is confirmed by the description of the secrets of their art by Columella (*Rust.* 9.8.10-12). Cicero's testimony (*Cato Maior de senectute* 57.7) speaks of slaves who collected honey in the forests. In the New Testament (*Math.* 3.4), the "wild honey" (obviously from trees and caverns) constituted the food of John the Baptist. Tree beekeeping is mentioned in Islamic beekeeping books of 15th c.[93]. The practice of collecting wild honey was reported for Cyprus as of 1365[94]. Tree beekeeping was widespread in Northern Europe, Africa and Asia from at least 8th c. up to the beginning of the 20th c.[95]. In Russia in the beginning of the 11th c. bee-woods of oaks are reported in the possession of aristocrats and monasteries[96]. In the Byzantine period this practice was so widespread that it necessitated special laws concerning who was eligible to gather honey from the tree cavities, depending on whether the tree was on a private property or not[97]. Bee-hunters existed up to recently in Greece[98].

Primitive honey collection concerned also honey-hunting in caves. Both Apollonius Rhodius (*Argon.* 2.131) and Virgil (*Aen.* 12.587-592) certify this practice. That the bees prefer to build their hives in caves was also known to Homer (*Od.* 13.102-12). When the 11th c. saint Lazarus Gelasiotis passed by Myra of Lycia, the locals asked him to pray for them so that they could reach in the dangerous declivities of a ravine in order to gather honey which existed there (obviously in cavities of trees or caverns)[99]. In the beginnings of the 15th century according to the testimony of a traveller, many caves of Crete entertained wild bees from which the locals gathered wild honey[100] explaining why many cave names in Crete derive from the bee[101]. The practice was used in China in the 3d c.[102]. Collection of

---

[93] Fahd 1968, 63
[94] Rizopoulou-Igoumenidou 2000, 391
[95] Crane 2000a, 127-136, 107-108, 221
[96] Crane 2000a, 129
[97] Kukules 1951, 355
[98] Loukopoulos 1983, 405; Faure 1999, 171
[99] Androudis 2000, 213
[100] Buondelmonti 1417, ch. XI; Faure 1999, 171
[101] Platakis 1977

honey from caves and rock cavities is mentioned in Islamic literature[103]. Evidence for a similar traditional practice exists for Lesbos and Achaia[104] while the same is reported for Italy, Serbia, Hungary, Albania, Croatia, and Cappadocia[105]. Wild bee nests however are rare nowadays in the Aegean world because of the mite *varroa*. In Minoan Crete, evidence[106] in favour of the existence of such primitive tree and cave beekeeping stems from literately sources of the historical period describing how Saturn was castrated by Rhea under an oak, where previously he had gathered and eaten a lot of honey (Kern *Orph. frag.* 154 and 189) and how honey-hunters tried to steal honey from the cave on mountain Dikty where Jupiter was given birth and nourished with wild honey (Boio *Ornithogonia* ap. Ant. Lib. 19).

Did a systematic (with beehives) apiculture exist in Bronze Age Aegean as in Egypt? To determine this we must turn to archaeological evidence.

---

[102] Crane 2000a, 118
[103] Fahd 1968, 63, 82
[104] Bikos 1996, 116, photo 14
[105] In Cappadocia, small cavities dug in soft rock were also used as beehives (Crane 2000a, 137). A similar practice was also the creation of beehives in the walls of houses as is testified for Cyprus in 1533 (Rizopoulou-Igoumenidou 2000, 393), India, Laos, Nepal, North Vietnam and elsewhere (Crane 2000a, 119-122, 125-6, 138-140)
[106] Grumach 1964

Beekeeping paraphernalia

Beehives

Davaras has shown that ideogram *168 ⌷ from Linear B, found exclusively in tablets from Knossos[107] along with the word *me-ri-te-o*, meaning "of honey"[108] depicts a prehistoric ceramic beehive[109]. The resemblance of the object depicted by the ideogram with the traditional horizontally placed pottery beehives of Crete ("solin" or "dipseli") and other Aegean islands (Antiparos, Paros, Anafi, Ios, Siphnos, Syros, Tenos, Ikaria, Karpathos, Lesbos) and Cyprus[110] is quite remarkable (fig. 4). A similarly shaped horizontal beehive was used in Morocco, Egypt, Israel, Jordan, Syria, Lebanon, Iraq, Iran and in the Arabic Peninsula[111]. Similar horizontal beehives dating to 10th-9th c. BCE have been recently discovered in Tel Rehov, Israel[112].

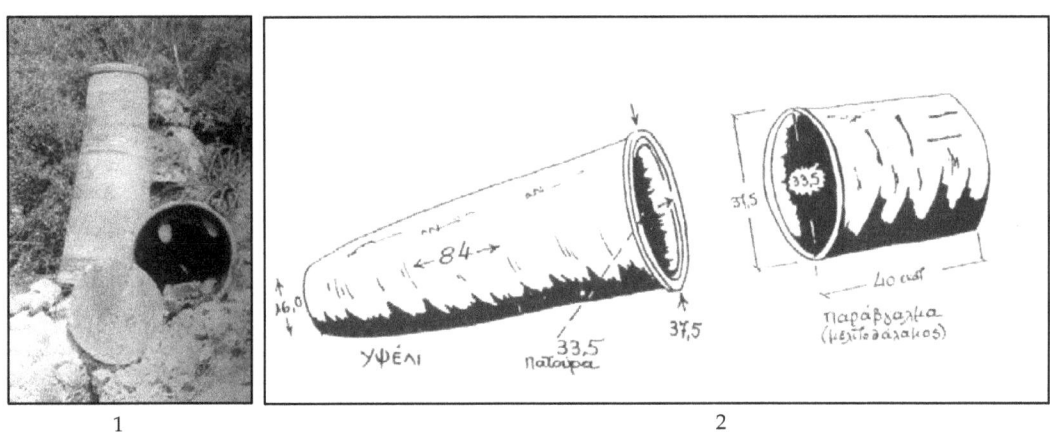

Fig. 4. (1)Traditional ceramic beehives from Crete (Rammou 2000, 428) (2) Traditional horizontal hive from Syros island with a deep extension ring (drawing Bikos 1994).

The horizontal beehive of Crete bears a strong resemblance to the prehistoric Egyptian beehive from which it might have evolved[113] (fig. 5). This is why it is believed that systematic (with beehives) apiculture spread in prehistoric times from Egypt to Crete[114] and to mainland Greece, a fact that as we have seen, is unanimously testified by the literary sources on the "mythology" of the ancient beekeeping.

---

[107] Tablets *Pp* and *U* 7505
[108] Duhoux 1975, where also bibliography for different opinions about the meaning of the ideogram
[109] Davaras 1986 and 1989. In the same author see bibliography for different opinions (double axe, adze or textile tool)
[110] Jones et al. 1973, Plate 85a, 85c, 85d; Crane 2000a, 193-5, fig. 22.3a and fig. 22.3b. In Cyprus a testimony of this type of beehive comes from travellers in 1801 (Rizopoulou-Igoumenidou 2000, 393)
[111] Crane 2000a, 167-8, 175, tb 21.4A181-2, fig. 21.6a
[112] Mazir and Panitz-Cohen 2007
[113] Jones et al. 1973, 403; Crane 2000a, 192; Lawall et al 2001, 177
[114] Crane 2000b

Fig. 5. Wall painting in tomb of Rekhmire, Luxor c 1450 BCE (Davies 1944 from Crane 2000a fig. 20.3b, 165). Horizontal beehives are depicted to the right

The Cretan traditional horizontal beehives are of 64-100 cm long, 29-40 cm in diameter at the front, tapering to a diameter of 19-23 cm at the back[115]. They are open at both ends, but a closed end is not unusual. Archaeological data indicate that the form of the Cretan and Aegean beehives has remained unaltered at least since the archaic period[116]; fragments of horizontal pottery beehives from 600 BCE to 400 CE have been excavated on six Aegean islands: Patroclus, Euboea, Makronisos, Kea, Delos and Chios[117]. Similar horizontal beehives of classic and later periods (including Byzantine) were found in Attica and Isthmia, and also in other Mediterranean regions, such as in Spain[118]. The traditional hives were positioned horizontally with one end sealed either with a wooden plate and mud, or a stone plate and clay. One or more small holes allowed the bees to fly out. The other end, which permitted harvesting of the beehive, was closed with a ceramic or wooden lid[119]. This was usually the back end, since bees store honey farthest from the flight entrance in order to protect it from marauders. In the interior of the hive little wooden bars were positioned across the walls of the beehive for the bees to build their combs, a practice mentioned already in the 12th c. *Book of Agriculture* by Ibn Al-'Awwam[120]. During harvesting, the lid was removed and bees were driven by smoke from the back end on to brood combs near the front of the hive. Hives with openings only at the front (as in recent times in some parts of Crete, mainland Greece and the Aegean[121]) require a more difficult harvesting procedure. A traditional practice, also known in antiquity, was to elongate horizontal hives by adding a bottomless cylindrical terra-cotta stem ("extension ring"), (fig. 4.2; fig. 6) which was fastened between the lid and the end of the hive which had projecting

---

[115] Crane 2000a, 192
[116] Francis 2000 and 2001; Hayes 1983; Homann-Wedeking 1950; Catling et al. 1981; Callaghan 1992; Di Vita 1993; Crane 2000a, 190-2
[117] Crane 2000a, 193
[118] Attica: Jones et al. 1973; Jones 1976; Jones 2000. A horizontal beehive, found in a farm villa in Attica dated to Hellenistic times, is in exposition in the museum of the Athens International Airport, *Eleutherios Venizelos*, (in the leaflet of museum *History and Culture of Messogias Attica*). Isthmia: Anderson-Stojanovic et al. 2002. Spain: Bonet et al. 1997. For a horizontal Byzantine beehive (6th – 7th c.) see Geny-Tsofopoulou 2002, 135, fig.147
[119] Crane 2000a, 192 and 387-8. Similar horizontal beehives with lids closing their ends were used till recently in Egypt (Kueny 1950, 88)
[120] Ibn Al-'Awwam 2000, 1022
[121] Crane 2000a, 210, tb. 24.4A

rims[122]. With this technique the beekeeper could easily separate the extension ring from the main hive and harvest part of its crop without disturbing the inner parts; this entailed using less smoke which was known to harm the taste of honey[123]. Additionally, the extra space provided in the hive prevented swarming.

Another type of ceramic beehive, the upright beehive, was in use in MMII Crete, Kassos and Karpathos[124]. These beehives, approximately 30 cm high and with an opening about 45 cm in diameter (fig. 7.1), resemble modern buckets. A Minoan upright beehive 18 cm high has also been reported[125]. The use of upright beehives in the 7th c. is documented for Cyprus while recent archaeological findings in Attica, Isthmia and Chios confirm the existence of upright beehives in the classical period[126] (fig. 7.2). The beehives from Isthmia have approximately the same dimensions as the ones from Minoan Crete[127].

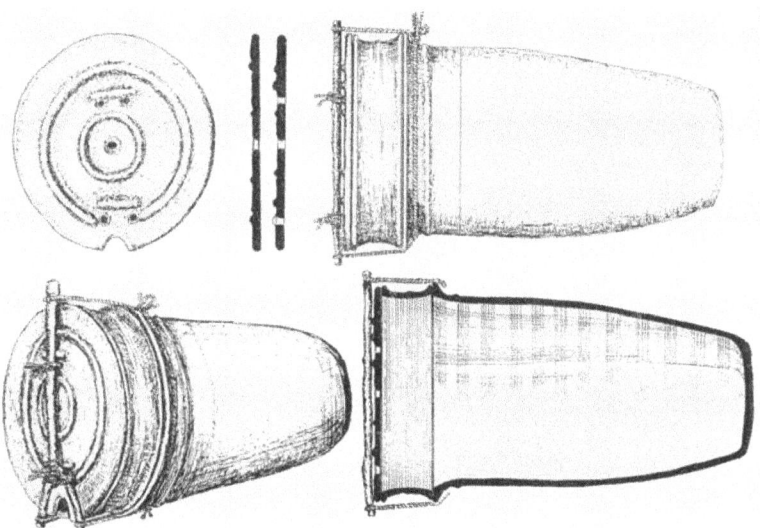

Fig. 6. Reconstruction of an ancient beehive from Trachones, Attica. The extension ring and lid are held together with a cord fastened around the projecting rim and a stick (Jones et al. 1973, fig. 19)

---

[122] Such a practice was widespread in various parts of the Mediterranean as in the Aegean, Malta, Morocco, Turkey, Lebanon but also in Iran, Iraq, Pakistan and India (Crane 2000a, 387-8). For the projecting rims of the beehives see Crane 2000a, 210

[123] Strabo's (9.399) and Lucianus' (*Navigium* 23.4) praise of the non-smoked honey is well known. In Byzantium too the non-smoked honey was highly appreciated (Kukules 1951, 354, n 2,3,4)

[124] Melas 1999, 487-488, Pl CVIIb, c, d, e, f and CVIIIa, c, d

[125] Melas 1999, 487, plate CVIIb

[126] Cyprus: Catling 1972 (reported by Crane 2000a, 184). Attica: Jones et al. 1973; Lawall et al. 2001. Ludorf (1998/1999) published a detailed typology of the beehive tracing its history from Classical to the late Roman period. Isthmia: Anderson-Stojanovic et al. 2002. Chios: Jones et al. 1973, 399, no 27

[127] The ideogram *212 from Knossos and Pylos could correspond to a vertical beehive

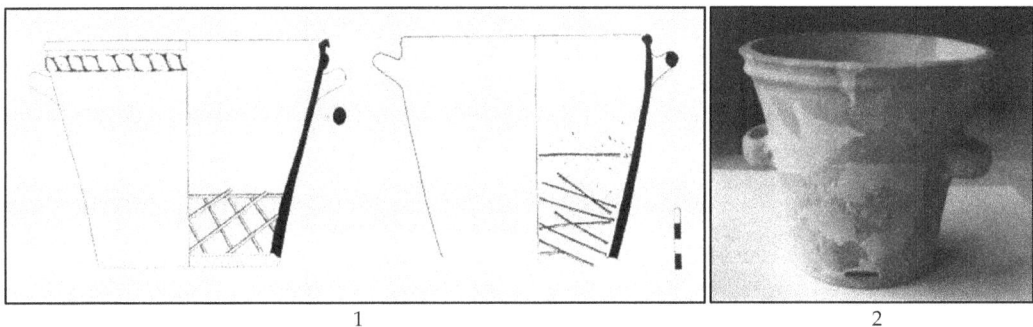

Fig. 7. (1) Minoan beehives (Melas 1999, plate CVIIb,c) (2) Ancient ceramic beehive from Isthmia (Anderson-Stojanovic 2002, fig. 9,)

Such vertical beehives were in use up to recently in Crete ("vraski") with a height of 36 to 41 cm and an orifice diameter of 35 to 41 cm[128]. The same type existed in Attica in the 17th c. ("anastomo kofini"), in the Cyclades in the 18th c. as in Syros and Kea ("ypseli") (fig. 8.1-2) and in Peloponnesus in Argos[129].

Fig. 8. (1, 2) Traditional ceramic beehives from Kea with top-bars (photos Bikos 1999)(3) A comb removed from a top-bar hive (photo P.Papadopoulou 1965, from Crane 2000a, 399, fig. 39.3a right)

The discovery of upright beehives in Minoan Crete provides important evidence for an early existence of beehives with "movable top-bars" reflecting an exceptionally high level of apiary knowledge[130]. Movable "top-bars" are wooden bars (traditionally made of wild laurel or other materials[131]) placed on the open end of an upright beehive. Bees can thus build their combs at these bars vertically into the beehive without any attachment to its walls. Such a set-up renders the removal of combs much easier and facilitates the apiarist in the continuous renewal of full bars with empty ones, thus increasing the production of

---

[128] Zymbragoudakis 1979; Crane 2000a, 397, tb 39.A; Nikolaidis 1955
[129] Attica: Wheler 1682; Cyclades: Rocca 1790, ii, 465-6; Argos: Efthimiou-Hatzilakou 1981-2
[130] The existence of prehistoric hives with movable top-bars is argued by Bikos 2000 and Crane 2000a, 404
[131] Bikos 1999, 5

honey[132] (fig. 8.2-3). This process can not be performed in horizontal beehives where the comb is fixed on internal bars positioned across the walls of the beehive. The wooden top-bars form the cover of the upright beehive along with cow dung, bunches of thyme and a slate plate. A hole in the base of the beehive allows the entry and exit of the bees.

Since upright beehives with movable top-bars permit the close observation of bee habits, Aristotle's detailed knowledge of apiculture is due, according to some scholars, to the existence of such beehives in his time[133]. The same argumentation can be used to explain the knowledge of the detailed bee habits depicted in the Mallia pendant. The upright beehive with movable top-bars is correctly considered by some authors as the forerunner of the modern beehive with "movable frames" which has played a very important role in the development of apiculture since its discovery in the 19th c.[134].

Before the wide distribution of the modern beehive (discovered in 1866 but not propagated in Greece until 1930), in no place did there exist only one type of beehive[135]. The coexistence of many types of beehives was a usual phenomenon in Roman times as well. Varro, Virgil, Columella, Pliny and Palladius mention the different materials used for beehives: biodegradable materials such as bark (Varro, Virgil, Columella, Pliny, Palladius), plant Ferula stems (Varro, Columella, Pliny, Palladius), woven wicker (Varro, Virgil, Columella, Pliny, Palladius, Hesychius s.v κυψέλη), hollowed log (Varro, Columella, Palladius), boards of wood (Columella, Palladius), cow dung (Columella)[136], and sun-dried mud (Columella) and other non biodegradable materials such as clay (Varro, Columella, Palladius), brick or stone (Columella)[137]. The existence of numerous types of beehives can be explained by the diversity of the environmental conditions, the availability of raw materials and different beekeeping practices.

A basic disadvantage of ceramic beehives is their weight and the difficulty that it poses of transporting them in order to extend the honey producing period by transferring bees to higher florescence regions. Areas at different altitudes or latitudes provide florescence at different seasons and those with different rainfall or soil support different bee-plant species. Migratory beekeeping (also called transhumance or pastoral beekeeping) was practiced either by land (transporting the hives with animals, like mules as recorded for Spain by Pliny *HN* 21.73-78[138]) or by sea (transporting the hives with boats). Celsus (ap. Columella

---

[132] See for a detailed description of the use of upright hives with movable bars in 17th c. Greece in Wheler 1682. Rocca (1790, ii, 465-6) referred to "more than one person" who described to him its use in Crete, and its uniqueness in the whole Levant

[133] Analytic arguments in favour and contra to this opinion exists in Jones et al. 1973, 405-8. Nonnus (*Dion.* 5.42) says that Aristaios discovered the "*beehives with many combs in line*" (πολυτρήτων στίχα σίμβλων); this phrase most likely reffers to beehives with movable top-bars

[134] Georgantas 1957; Ifantidis 1983; Bikos 1998; Crane 2000a, 457-460; Protopsaltis 2000. Beehives with movable top-bars existed also in N. Vietnam at least since the 19^{the} c. (Crane 2000a, 400-2, fig. 49.4a)

[135] As was the case in Crete (Rammou 2000, 428-430; Nixon 2000) and elsewhere in Greece (Liakos 1999a; Rammou 2000; Graham 1975, 75; Anderson - Stojanovic 2002, 366, n 34)

[136] As with some traditional hives in Greece (Loukopoulos 1983, 392; Katsouleas 2000, 358)

[137] See Crane 2000a, 203, tb 24.1A. Hesychius reports six different names for beehives which probably indicates different forms and materials

[138] By the late 1800s, trains were used, while during the 1900s road vehicles of various types and sizes performed this task (Crane 2000a, 347)

Rust. 9.14.20) explained the general principles and precautions of transporting hives. Migratory beekeeping with donkeys or by boats was practiced in 3d c. BCE Egypt: beehives were placed on boats that sailed along the Nile in search of regions with florescence[139]. The same practice was recorded in Egypt almost two thousand years later (in 1740)[140]. Celsus (ap. Columella *Rust.* 9.14.19) records the migratory beekeeping that was practiced in Greece (Peloponnesus, Attica, and Euboea) and in Sicily (Hybla). Columella (*Rust.* 9.14.19) also reports migratory beekeeping with boats from Cyclades to Skyros. Migratory beekeeping with boats was a widespread apiarian practice until recently[141]. In 1790 Rocca recorded the transportation of beehives along the coasts of Asia Minor[142]. Beehives from Arnaia in Chalkidike were transferred to Mount Athos and to Thasos in springtime[143]. Also in Chalkidike, until 1960, small boats loaded with beehives were circumnavigating the gulfs[144]. In Ios, Cyclades, they transported the beehives with fishing boats[145]. Similar accounts exist also for France, Belgium, China and Japan[146]. In China the boats transporting the beehives had marks on their hull in order to indicate the increase of draught due to the increase of weight from honey accumulated in the beehives during the voyage. Precisely the same strategy is described by Pliny (*HN* 21.43) in Hostilia in Italy, where Roman apiarists loaded their beehives in boats and travelled along the Po river to exploit the rich florescence. Migratory beekeeping with boats along rivers is also reported in America and in Romania[147]. Perhaps the oldest written testimony of migratory beekeeping by boat is the story reported by Philostratus (*Imag.* 2.8.6) and Himerius (*Or.* 59.5) that the boats of settlers of Androklos, son of Kodros (who crossed the Aegean and reached Ephesus), were led by bees. That the Minoans transported their beehives by boats can be deduced by the discovery of a pottery boat model (MMI) carrying honeycombs in its cargo hull[148] (fig. 9). Because it was found in a grave it was interpreted as a symbol of "after death voyage" but its purpose could be simply to denote the activities of the occupant of the grave during his lifetime.

---

[139] Newberry 1938. For the transportation of beehives by land, see *P. Cair Zen.* III 59467 (*SB* 6989)

[140] Maillet 1735. *Description de l'Egypte.* 2 vol. Paris, 24 ; Pococke 1743. *Description of the East* I, 210; Savary 1787. *Letters on Egypt* 2nd ed. II, 207 (all reported by Newberry 1938 who certifies that he observed in the Nile, in 1890, the same practice). An interesting detail that Savary reports is that all the beehives had the mark of their owner in order not to get mixed up with others.

[141] Crane 2000a, 347-352. For Greece see Typaldos-Xydias 1927

[142] Rocca 1790

[143] Gaitanou-Giannou A. (unpublished notes 1930) in Kyrou 2000, 377. See also Eckert 1943 (reported by Petropoulos 1957, 356)

[144] Papagelos 2000, 199

[145] Rammou 2000, 423

[146] France: Cotton 1842, 338 (reported by Ransome 1937, 226); Belgium: Zeghers M. 1780. *Mémoires sur les questions proposes par l'Académie Impériale... de Bruxelles... en 1779.* Brussels (reported by Crane 2000a, 349); China and Japan: Marchenay 1979 (reported by Crane 2000a, 349)

[147] Crane 2000a, 349 with references

[148] Davaras 1984, tb. 6a-b, fig. 1

Fig. 9. Minoan ceramic model of a boat transporting honey combs. Mitsotaki collection (Ίδρυμα Ν. Γουλανδρή, Μουσείο Κυκλαδικής Τέχνης, Αθήνα 1992, 106, photo G. Giannelos)

Hives most suitable for migratory beekeeping were sturdy light by weight, such as those made of wooden boards or the woven wicker beehives[149]. A bell-like wicker beehive (skep) was widespread up to recently in Greece, especially in the Chalkidike peninsula and in Crete[150] ("epistomo kofini"), in Europe and other parts of the world[151] (fig. 10.1-3).

Fig. 10. (1) Traditional knitted beehives from Florina (Rammou 2000, 434) (2) Skeps from 15th c. Ibn Butlân, Taqwim es Siha. Allemagne, Rhénanie (3) Traditional skep from France (photo from Richards-Mantzoulinou 1979, fig.2) (4) The omphalos of Delphi

However, its existence in Greece since antiquity has been questioned and it has been suggested that the skep came to the Mediterranean in the 12th c. from Northern Europe[152]. Nevertheless, a skep appears in a mosaic of the 6th c. in Jordan (Madaba) depicting the fourteenth *Idyll* of Theocritus with bees stinging Eros while he steals honey from a knitted

---

[149] Crane 2000a, 219; Georgandas 1957; Rammou 2000, 430
[150] Leontidis 1986, 40
[151] Crane 2000a, 219-21, 232-6, 241-57, 265
[152] Crane 2000a, 183, 219

beehive[153]. The description of Petronius (*Sat.* 39.14) of a round as an egg (*quasi ovum corrondutata*) beehive and the account of Virgil (*G.* 4.33) of a beehive "*woven of pliant osier*" constitute, in our opinion, the main corroborations for the existence of skeps in Roman times[154].

It has been proposed that the omphalos of Delphi (fig. 10.4) and that of other oracles[155] were beehives, given their resemblance with the traditional skeps[156]. Characteristic traits of the omphalos from Delphi are the embossed lines. Their meaning has been interpreted as either representing necklaces or a net[157]. We favour the opinion that they represent the knitted material of a skep[158]. Ideograms *134 and *190 of Linear B, on tablets from Mycenae[159], are dome-shaped with three horizontal dashes crossing or flanking each side of the dome. An analysis of some of the words associated with the ideograms shows that they relate to words that mean "made of wicker" and "conical baskets". This could lead to a safe identification of the two ideograms with knitted beehives[160]. An omphalos-like object resembling that of Delphi is depicted on a fresco from Minoan Knossos[161]. We believe that an omphalos-like knitted beehive can also be recognised at symbol 7 "◌" from the disk of Phaistos (which Evans identifies with a "woman's breast")[162], as well as at the ideogram *179 of Linear B[163].

But probably the most important testimony, unnoticed until now, for the existence of the knitted beehive in the East Mediterranean since prehistory, is the vessel on the left hand of the figure from the wall painting of tomb 101 in Luxor (fig. 11). This we interpret as a skep with combs, been transported upside down with its opening covered to prevent bees escaping, as was the practice with skeps up until recently (fig. 12)[164].

---

[153] Piccirillo 1993

[154] We believe that the clay models of "granaries" found in the Geometric temple of Ano Mazaraki (Rakita) in Achaia, now in display in Patras Archaeological Museum (Petropoulos 2002, 143-164, Tav. 3, fig.4), could represent wicker skeps similar to the ones used by traditional beekeepers in Attica and in France (see photos in Rammou 2000, 431 and Crane 2000a, 247, fig. 27.3a). We are grateful to Sarah P. Morris for drawing this to our attention. The beehives mentioned in the Attic Stelai (*IGI3*, 426,56) were considered by Pritchett (1956, 260) as been made of wicker

[155] See the omphalos at the statues of Asklepios (Vatican Museum, Braccio Nuovo, Museo Archeologico Nazionale, Naples). See also Richards-Mantzoulinou 1979, 84-85 for other locations as in Eleusis

[156] Zafiropulo 1966, 40; Richards-Mantzoulinou 1979

[157] Cook 1895, 5 n 32

[158] Richards-Mantzoulinou 1979, n 66. We have already concluded an analysis that the "myth" about the four temples of Delphi, as reported by Pausanias and others, can be interpreted in a beekeeping perspective. We will present our analysis elsewhere

[159] MY *Oi* 701; MY *Go* 610

[160] Melas 1999, 489. For evidence for the existence of basketry and weaving in Greece since Early Neolithic see Perles 2002, 243-6

[161] Evans 1921-1935, II, 839-40, fig.555

[162] Evans 1921-1935, I, 651 n1. Davaras (1986, 40 n.13) mentions the opinion of L. Pomerance that the Phaistos Disc sign 24, the so-called "Lydian tomb", is a knitted beehive

[163] Proposed as a likely depiction of an omphalos-like knitted beehive by P. Faure (Vandenabeele 1979, 24, reported by Davaras 1986). According to the *Homeric Hymn to Apollo* (388-544) the first priests of Delphi were Cretans from "*Minos' Knossos*"

[164] Crane 2000a, 219

All of the above suggest that the knitted beehive existed in the Mediterranean world since prehistoric times.

Fig. 11. *Wall painting from tomb 101 in Luxor showing on the left hand of the male figure a knitted beehive full of combs (photo: The Egyptian Museum from Crane 200a, 596, fig. 54.3a).*

Fig. 12. *A skep with honeycombs from modern Macedonia, Greece (photo A.Harman from Crane 2000a, 220, fig. 25.4b)*

Smoking pots

Ancient and Byzantine apiarists smoked their bees in order to pacify them (Pl. *Phdr* 91 C; Arist. *Hist. an.* 623b; Plin. *HN* 11.15.45; Verg. *G.* 4.228; *Geoponica* 15.5, 15.6), just as modern ones do[165]. This practice is depicted on a painting, dating to 2400 BCE[166], from an Egyptian temple (where horizontal beehives are present as well) and in the Egyptian grave No 100 of Rekhmire in Luxor, of 1450 BCE[167] (fig. 5). The most primitive technique of smoking bees was with torches, a practice used until recently in certain regions of Greece[168]. However, smoking pots of a particular shape are needed in order not to burn the bees or the beehives (made of wood or wicker), and to be able to direct the smoke more accurately[169]. The basic typology of Roman smoking pots is given by Columella (*Rust.* 9.15.5): they should be ceramic, with a handhold, shaped like a narrow pot with two big apertures. The beekeeper should blow on the fuel (usually dried cow dung) in the pot through one aperture so that the smoke could emerged from another. The same basic principle governed the manufacture of smoking pots up to modern times, as can be seen in pictures of traditional smoking pots from Greece and elsewhere (fig. 13).

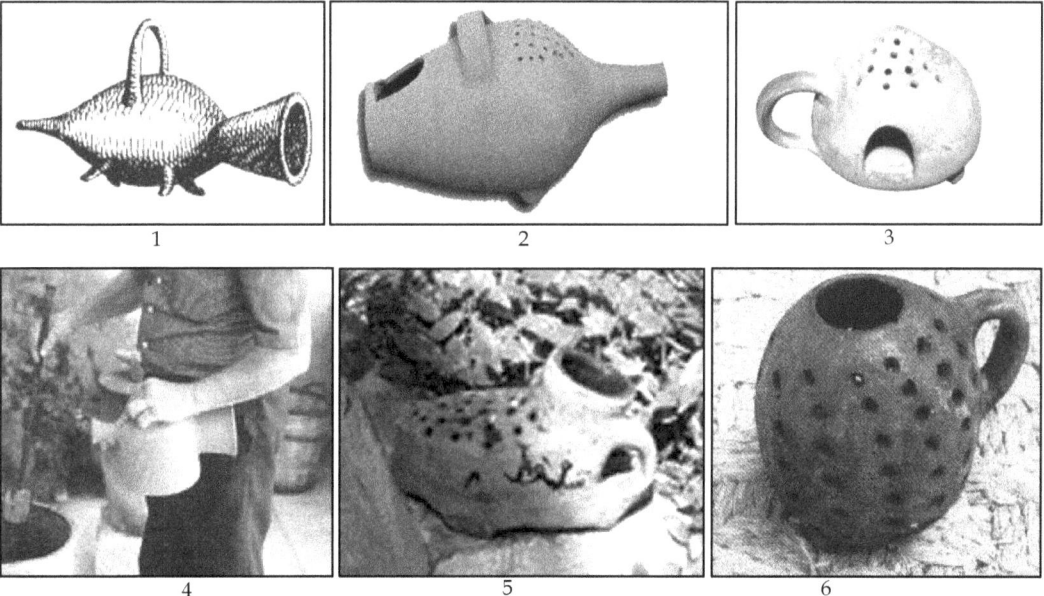

Fig. 13. Types of traditional smoking pots (1) Greece (Rocca 1790) (2) Lesbos (Koutri 1999, 320, fig.) (3) Lesbos (Gianopoulou 1998, 67) (4) Crete (photo E. Crane from Crane 2000a, fig 34.2c) (5) Tunisia (photo http://www.ppru.cornell.edu) (6) Algeria (photo I. Ritchie from Crane 2000a, 342, fig. 34.2b)

---

[165] Kukules 1951, 354
[166] Crane 2000a, 163-164, fig. 20.3a
[167] Davies 1944 (reported by Crane 2000a, 164, fig 20.3b)
[168] Loukopouos 1983, 398. For the same practice in other countries see Crane 2000a, 54, 59, 341
[169] Crane 2000a, 341

It appears however, that this design existed already in prehistoric times, since similar Early Helladic smoking pots were found in Northern Greece (fig. 14) and in Peloponnesus (Olympia)[170].

Fig. 14. Earyl Helladic smoking pot from Archontiko, Macedonia (photo PapaefthimiouPapanthimou 1997, fig.11)

The smoking pot from Olympia is similar to that used in traditional beekeeping in Kythnos, Paros, Naxos and Crete[171]. Tsountas published in 1908 a perforated ceramic vessel from the Final Neolithic of Sesklo in Thessaly which he identified it as a smoking pot for bees[172]; it, indeed, closely resembles the Early Helladic smoking pot from Macedonia. This important discovery, as far as we know, constitutes the world's oldest apiculture vessel and its existence implies organised apiculture in Neolithic Greece. A Late Minoan beekeeping smoker was found in a cavern in the gorge near the Zakros palace[173] and a similar one in the "oikia I"[174] of the palace (fig. 15). Both have obvious typological similarities with the traditional smoking pot of Greece (Lesbos) and Tunisia (fig. 13.2; fig. 13.5).

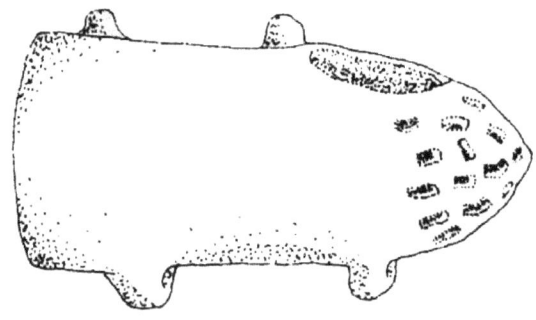

Fig. 15. Smoking pot from Zakro (Davaras 1987)

---

[170] N. Greece: Papaefthmiou-Papanthimou 1994, 8:90, photo 11; Papaefthmiou-Papanthimou 1997; Papaefthmiou-Papanthimou 1998, 122:855, fig. 163. Olympia: Rambach 2002, 194, fig. 28
[171] Rambach 2002, 194, fig. 29, n 114 with references
[172] Tsountas 1908, 274, fig. 198
[173] Davaras 1989, 3 and Platon 2005, 57
[174] Davaras 1989. Davaras also recognizes as beehive smoking pots similar objects found in Knossos, Phaistos, Agia Irene and Cyprus (Egomi)

Vessels from the "Snake Room" in Knossos

The smoking pot found in "oikia I" of Zakros "palace" led Davaras to call this particular oikia, a beekeeper's residence[175]. We believe that this is not an isolated example and we

Fig. 16. Vessels from "a private house" of Knossos (Evans 1921-1935, IV, 95, fig. 109)

will show below that a similar oikia existed in Knossos, a much more rich one in findings, but not been recognized as such before. The house in question is a private house, located southwest of the "South-West Treasury House" of the palace of Knossos. A number of vessels were found in a small room of this house. Evans called the particular room the "snake room" and dated the vessels to MMIIIb-LMII[176] (fig. 16). This little room opened on to a passage-way which was however only partially preserved. By the entrance of the little room stood a large jar, 71cm in height and about 30cm wide, which was a repository of what appears to have been a complete set of clay vessels and other utensils. Both the jar itself and its contents had been much broken. Among the vessels in the jar some are perforated (No 1, 2, 3, 10 in fig. 16). One of them (No 2 in fig.16; fig.17.1) has two large openings on the sides and many small holes. Due to its snake-like handles, it is generally

---

[175] Davaras 1989
[176] Evans 1921-1935, IV, 155-156, fig. 118, 119

identified as paraphernalia for a snake cult[177]. But it could have been, instead, a smoking pot since it has many features in common with other smoking pots. Another perforated vessel (No 3 in fig. 16; fig. 17.2) has only one opening at the top and many small holes. It is probably a smoking pot too, but of a type encountered in the traditional apiculture of Algeria[178] (fig. 13.6).

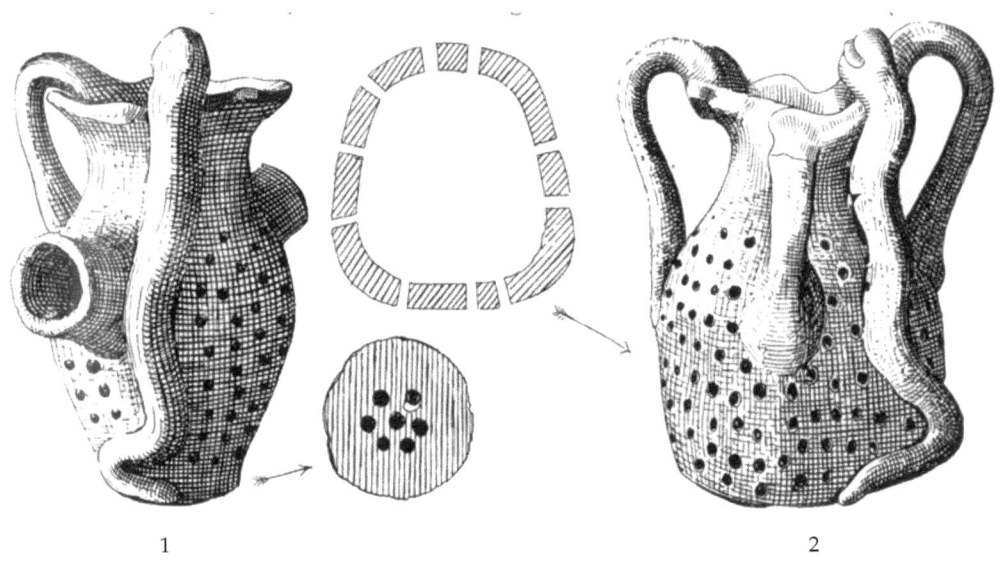

Fig. 17. *The vessel No 2 (**1**) and No 3 (**2**) from the private house of Knossos (Evans 1921-1935, IV, 155, fig. 119)*

A similar smoking pot for beekeeping was found in the prehistoric settlement (EH/MH) of Palamari in Skyros (fig. 18). We should not be surprised by the use of different types of smoking pot within the same region, since such practice is not uncommon. It is reported that in Crete, in 1985, six or seven different styles of smoking pots were in use[179].

Fig. 18. *Smoking pot from Palamari Skyros EH/MH (Skyros Museum. The findings of Palamari. Hellenic Ministry of Culture. 11th Ephorate of Classical Antiquities. 2005, photo 4).*

Another utensil (No 8 in fig.16) found in the jar is a circular object (diameter 25 cm), divided in four parts by four channels and standing on three legs (fig.19). We consider

---

[177] Nilsson 1950, 90
[178] For a photo see Crane 2000a, 342, fig 34.2b
[179] Crane 2000a, 342

Evans' hypothesis of a vessel for food offering to snakes ("snake table")[180], to be improbable, and certainly unprovable.

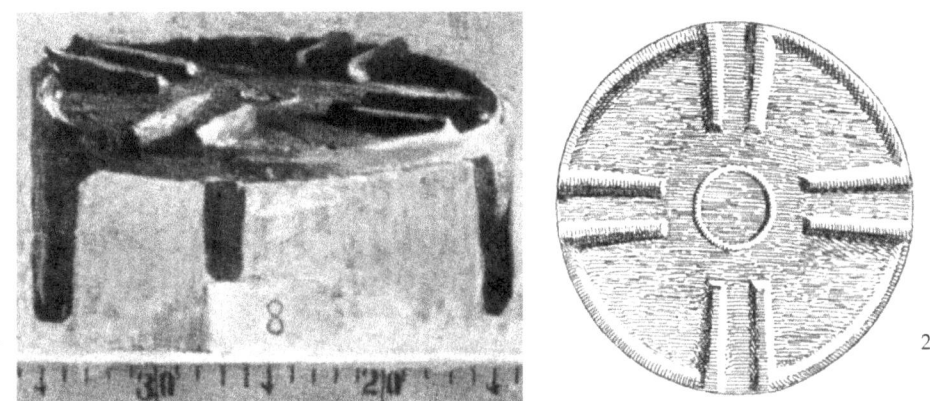

Fig. 19. (1) The object No 8 (modified from Evans 1921-1935, IV, 95, fig. 109) (2) the same vessel in ground plan (Evans 1921-1935, IV, 149, fig. 115a)

It could, however, be a honeycomb press (fig. 20). Combs could have been placed in the four compartments between the channels and then manually pressed with a wooden board (not preserved). Pressure would result in honey escaping through the four channels and flow into vessels (or a big dish) placed below the edge of each channel (such vessels could be the jags No 18, 19, 20 and 22 in fig.16 that Evans calls milk-jags for snake offerings).

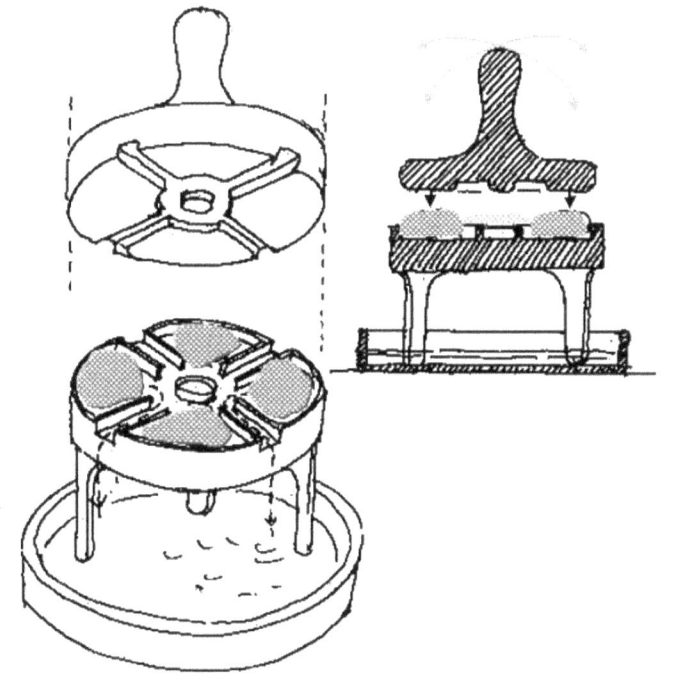

Fig. 20. The function of vessel No 8 for honey segregation (drawing V.A. Harissis)

---

[180]Evans 1921-1935, IV, I, 149, fig 115b; Nilsson 1950, 90

A press with channels for the flow of honey was used by traditional beekeepers in Cyprus[181] and in Greece[182] (fig. 21-22).

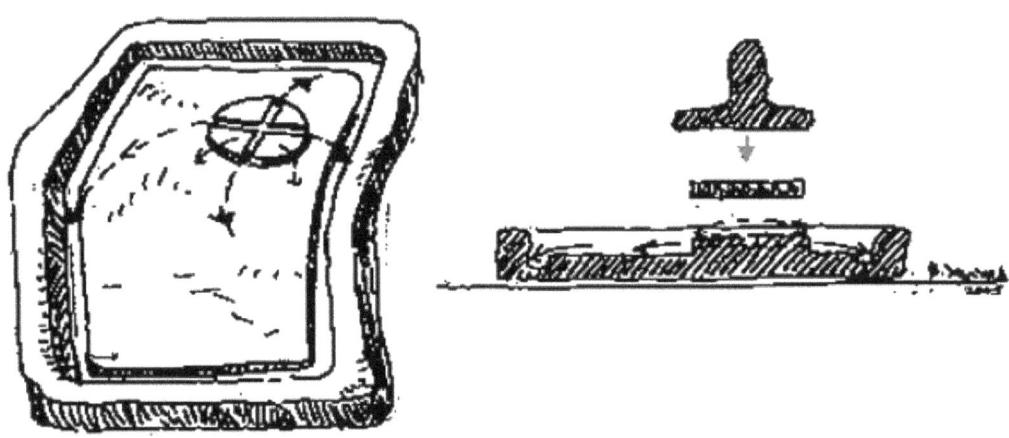

Fig. 21. Basin for pressing honey combs from Cyprus and its working principle (drawing V.A. Harissis based on a photo in Rizopoulou-Igoumenidou 2000, 404)

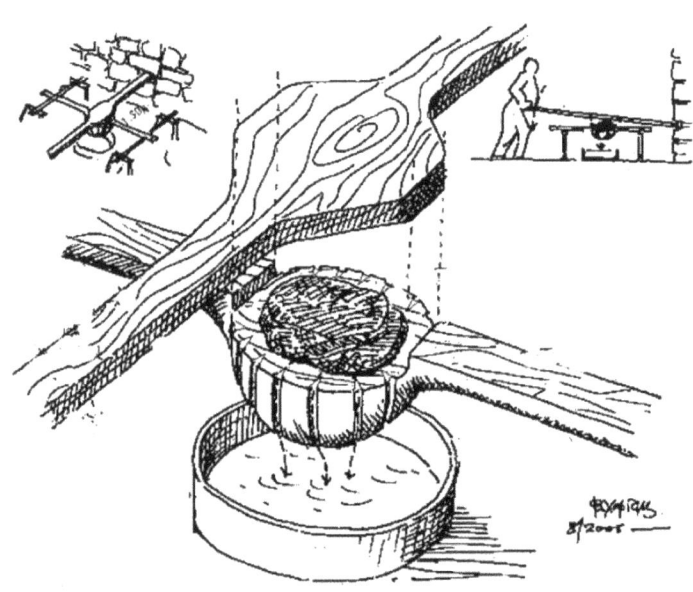

Fig. 22. Traditional honey extractor as described by Loukopoulos 1983, 400-1 (drawing V. A. Harissis)

---

[181] We saw a similar Hellenistic press (characterized as an olive press) in the Archaeological Museum of Setia. Nikolaidis (2000, 135) reports that the traditional comb presses were similar to those for grapes.

[182] Loukopoulos 1983, 400-1, fig. 53. The simpliest vessel for wax extraction from the comb was a ceramic stainer into which the comb was placed and with manulal pression the honey was separated from the wax (see for example figure 46.1b in Crane 2000a, 483). For such a Neolithic perforated vessel from the Nothern Aegean see Decavalla 2007 (we owe to S. Morris the indication of this information and reference).

The three "cylinders" or "tubes" (height 28 cm and exterior diameter of base 9.6 cm) (No 4, 5, 6 of fig.16; fig. 23) found in the "snake room", have two pairs of cups, symmetrically attached to their sides. Evans suggested that these cups were "made to contain some kind of drink offering to snakes" and labelled them "cylindrical snake vessels"[183]. We believe, however, that the cups were used as receptors of the overflowed liquid content of the tube. More specifically, we propose that these vessels served as wax extractors from the combs once honey was extracted first[184].

Fig. 23. Cylindrical vessel from the "snake room" of Knossos (photo Herakleion Museum)

The extraction of wax from the remainder elements of the comb (pollen, brood) is achieved, as Pliny (*HN* 21.83-84) and Columella (*Rust.* 9.16.1) recommend, with the use of boiled water.

---

[183] Evans 1921-1935, IV, 142, fig 111; Nilsson 1950, 90

[184] In a perforated dish from Knossos, Faure (1999, 171-2) recognizes a honey extractor. He compares it with similar objects from Troy and Neolithic Switzerland. By putting the comb in the vessel and by applying pressure, the honey spilled from the holes while the wax remained in the vessel. Melas (1999, Plate CVIIIe) presents a completely different conical vessel which he considers to be a honey extractor. The vessel No 3 in fig 16 from the Knossos Snake Room (fig. 17.2) could be a vessel to segregate honey from wax as the one used by traditional apiarists in Poland (for a photo see Crane 2000a, 483, fig. 46.1b). But the small diameter of the opening (insufficient for placing the combs) argues against this hypothesis. Traditional beekeepers also used to place the combs inside a simple linen sac. By applying pressure on the sac the honey flew out of the sac and was separated by the other comb components that remained in the sac.

The wax being lighter than the other comb components floats in boiling water and is collected from the surface. The same principle was used by traditional beekeepers in Greece[185]. Thus, we suppose that combs were placed in these Minoan containers and then the vessel was filled with boiled water. The heating of the water was done probably by placing little water jags (such as No 9, and 23 in fig. 16) over a fire alight in the vessel No 7 in fig. 16 which had traces of ashes. Filling with boiled water the tube forced the molten wax to rise to the surface, and by deliberately overflowing the container the wax was gathered in the cups[186]. The wax, after cooling, was removed from the cups, having taken their hemispherical form. The form and the diameter of the cups resemble the traditional and Byzantine vessels, used for the same purpose[187] (fig. 24). Based on the same principle (molten wax rising to the surface of boiling water) two metallic wax extractors, the "Gerster extractor" (fig. 25) and the "Mountain Gray Extractor", were in use in the 19th and 20th c. respectively[188].

Fig. 24. Traditional vessel for beeswax (Vrontis 1939, 236)

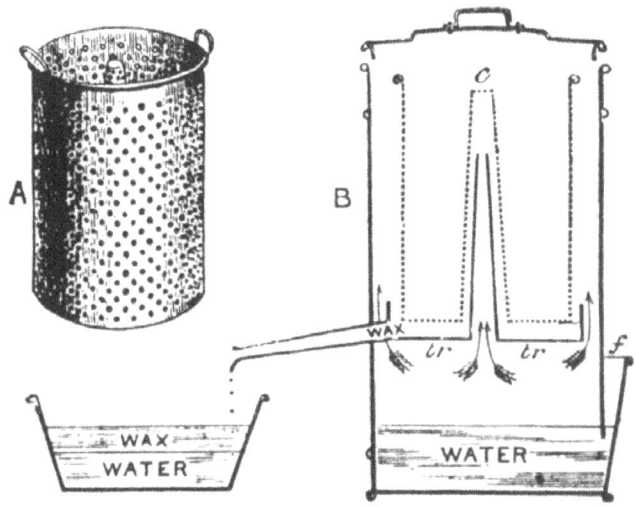

Fig. 25. The "Gerster wax extractor" (Cheshire 1888). *(A) Depicts the strainer pot where the combs are put and corresponds to the main body of the cylindrical Minoan vessel. The pot, is placed on a tray (**tr**) which was filled (through **f**) with boiled water passing through the opening of the main cone (**c**) into the comb strainer. Through the holes of the strainer the wax and the water flow in a dish just like the cups of the Minoan cylindrical vessel.*

---

[185] Liakos 1996, 371-2
[186] A similar practice was used, traditionally, by the apiarists in Cyprus (Filotheou 1980; Rizopoulou-Igoumenidou 2000, 404)
[187] Vrontis 1939, 206. These wax cups are called "*kyparia*" in Chalkidike and in Paros (Papagelos 2000, 198)
[188] Crane 2000a, 497, fig. 46.7d

Some other vessels (No 11, 12, 15, 16 in fig. 16; fig. 26) from the same room resemble the dish depicted in the mural from Egypt containing honey combs (fig. 11, fig. 27). This dish, in turn, resembles the traditional comb-dish from Casmir (fig. 28)[189] and the two dishes, one on top of the other, that can be seen on the wall painting from the tomb of Rekhmire, sealed with mud and containing combs (fig. 5; fig. 29). A similar dish with traces of a comb was found in a tomb in Deir-el-Medina, West Bank in Upper Egypt, c. 1350 BCE[190].

Fig. 26. Comb dishes from the upper shelf of the "snake room" in Knossos (modified from Evans 1921-1935, IV, 95, fig. 109).

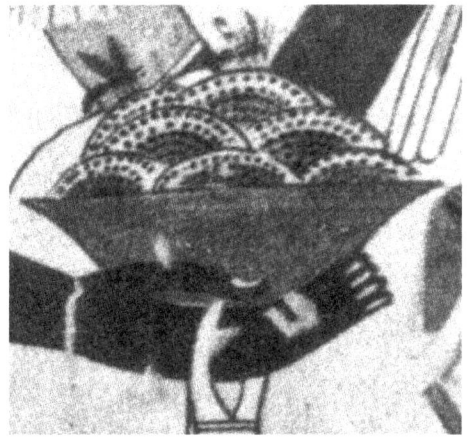 

Fig. 27. Comb dish depicted on a wall painting of tomb 101 in Luxor (detail from Crane 2000a, 596, fig. 54.3a).

Fig. 28. Traditional comb dishes from Kashmir (India) (photo E. Crane from Crane 2000a, 282, fig. 29.5d).

---

[189] By Crane 2000a, 165
[190] Crane 2000a, 166, fig. 20.3d

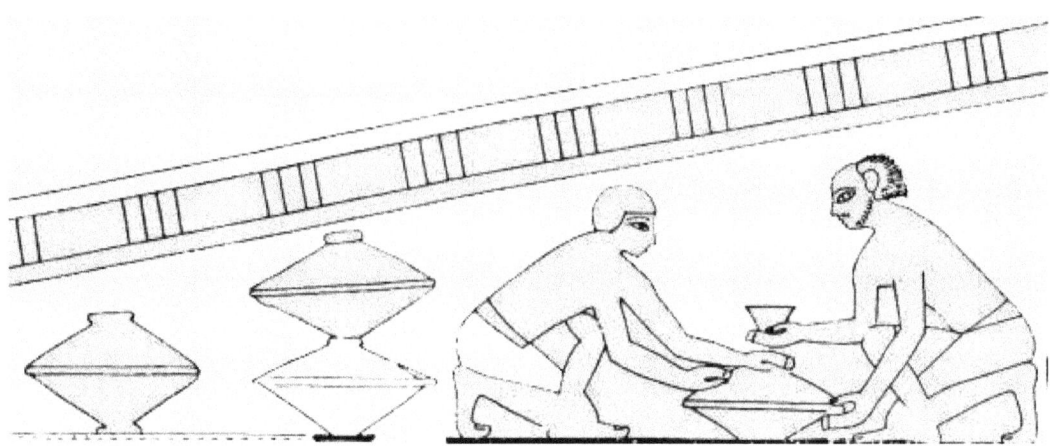

Fig. 29. Comb dishes depicted on the wall painting of tomb 100 of Rekhmire in Luxor (detail from Davies 1944)

The object No 1 in fig. 16 has been identified by Evans[191] as "three sections of a naturally formed wild honeycomb with a snake coiling round the vessel with a grub in its mouth" (fig. 30). We propose an alternative interpretation that of a rather sophisticated wasp trap - wasps being the worst enemy of bees in Southern Greece and the Aegean islands[192]. We believe that the vessel was deliberately made to resemble a honeycomb in order to "befool" the wasps enter the vessel. Possible an immersion in honey-water (*hydromelo*) would render it a more efficient trap. The vessel was probably placed near the beehives and when several wasps were trapped inside, the beekeeper threw it into the fire or into the water thus burning or drowning the wasps. Aristotle (*Hist. An.* 627b) reports a similar way for attracting wasps with a piece of meat placed in a dish and then killing them by throwing the meat into the fire. Finally, it should be stressed that vessel No 14 in fig. 16 is probably an upright beehive. Its resemblance to the Minoan hive of figure 7.1 is remarkable, and the two have very similar dimensions.

Fig. 30. The perforated vessel No.1 from the "snake room" (photo Herakleion Museum)

The existence of a beehive, smoking pots, honey extractor, wax extractors and comb-dishes in this "snake room", suggest that it is an apiarist room and not a room associated with a snake cult. If it is true, as it is generally accepted, that the "snake room" in Knossos and the "oikia I" in Zakros were attached to the "palace", we could suppose that in Crete, as in Egypt, apiculture was controlled by beekeeping overseers. Elaborated jewels, as the Mallia pendant, worn by these officials could represent emblems of their

---

[191] Evans 1921-1935, IV, 154-5, fig. 118a,b
[192] Reras 2001, 24

status and the numerous seals bearing a bee sign could have been used by them for the secure turnover of the valuable beekeeping products like the Near Eastern and Egyptian contemporaneous examples. We believe that we should also identify with beekeeping overseers the owners of some of the most famous rings, seals and sealings depicting "cult" scenes since detailed apiculture scenes should be recognized instead. This, we will present in the second part.

Part 2: The iconographical evidence.

Minoan and Mycenaean symbols revisited

Contra Evans

Glyptic Aegean Bronze Age scenes with their specific technical characteristics and miniature sizes pose difficult interpretation problems. In interpreting iconic forms one can not expect to assign singular and universal values as would be the case with alphanumeric characters. Images are polysemous and highly ambiguous. The range of acceptable alternatives although restricted by the interrelation of the forms represented in a scene, is not confined to a single solution. In order to assign an interpretation to a certain form in an image, two prerequisites are needed: a reference (a memory of a similar form) and the context in which it occurs.

A form may suggest several hypotheses when seen in isolation but only one in a certain context: a form of bird in a scene with a Christian context represents the "Holy Spirit" but in a scene with a naturalistic context represents a pigeon. Even if a certain form bears little physical resemblance to a particular object, context convention permits this form to be consistently recognized as this particular object. Thus, all meaning is contextually bound. But context in its turn is culturally bound. Our cultural tradition creates certain contexts, "mental sets", which affect significantly our perception and "deciphering" of works of art[193]. If the viewer of an image brings uncritically into play his own cultural assumptions then he is likely to deploy his own assumptions and expectations to make sense of images to which those assumptions could be alien, thus distorting their true meaning. This could be especially true for images produced by a civilization that flourished more than three thousands years ago as the prehistoric Aegean scenes depicted on signet rings. Some particular "peopled" scenes, comprising a small percentage of the Minoan and Mycenaean corpus, have inspired generations of scholars - all deeply influenced by the work of Evans - to explain aspects of Aegean Bronze Age religion on the basis of their imagery. They feature principally female figures that appear engrossed in actions and situations that take place in nature (fig. 31). What led Evans to deploy a religion context to these scenes in the first place were his own cultural assumptions embedded in the intellectual trends of his time. One of Evans' major influences[194] was the 1861 work of Johann Jacob Bachofen *Mother Right: An Investigation of the Religious and Juridical Character of Matriarchy in the Ancient World* that theorized a matriarchal phase in the

---

[193] Gombrich 1977, 53-78; Eco 1976, 204-5, Sourvinou-Inwood 1989, 241-2
[194] Lapatin 2003, 68-72

development of all culture. Bachofen linked the advent of matriarchy to the birth of agriculture in the belief that early (prehistoric) peoples throughout the world correlated the ability of women to engender new life with that of the earth. He and others, then, posited a universal cult with a female deity associated with earth. The Romantic Movement developed further the social context of this idea: in their matriarchal world, prehistoric humans peacefully coexisted with nature as well as with one another in a loving, egalitarian society protected by their Mother Earth Goddess. Bachofen and his followers found evidence for these developments in the myths of Anatolia, India, and central Asia as well as of Italy and Greece. Evans immediately and without hesitation, recognized the female figure on the rings and seals as the Mother Earth Goddess, an all-powerful female deity of the Minoans. But even though Bachofen applied evolutionary theories to the development of culture in a manner that is no longer considered valid (modern archaeology and literary analysis have invalidated many aspects of his historical conclusions) Evans' interpretation still dominates.

If however, we should question the religious context of Evans and others on the above basis what safe ground is there left to proceed? There is one unquestionable fact: many of the so called religious scenes have a strong naturalistic element. Could it be then that the scenes represent some rural practice instead of the supposed religious practices? This working hypothesis leads us to a much more solid and tangible ground since agriculture practices in the Eastern Mediterranean and Greece - as opposed to religious practices - have remained practically unaltered until relatively recent times. In this agricultural context the use of ethnographic sources along with the archaeological and literally evidence can function as "critical correctives to our imagination"[195]. By applying a similar method in the first part of this monograph we have interpreted some of the supposed cult vessels in Knossos as simple apiculture vessels. In the following we will attempt to analyze the forms appearing on some Minoan and Mycenaean rings, seals and clay sealings, in this new context.

---

[195] Morris 1992, 210

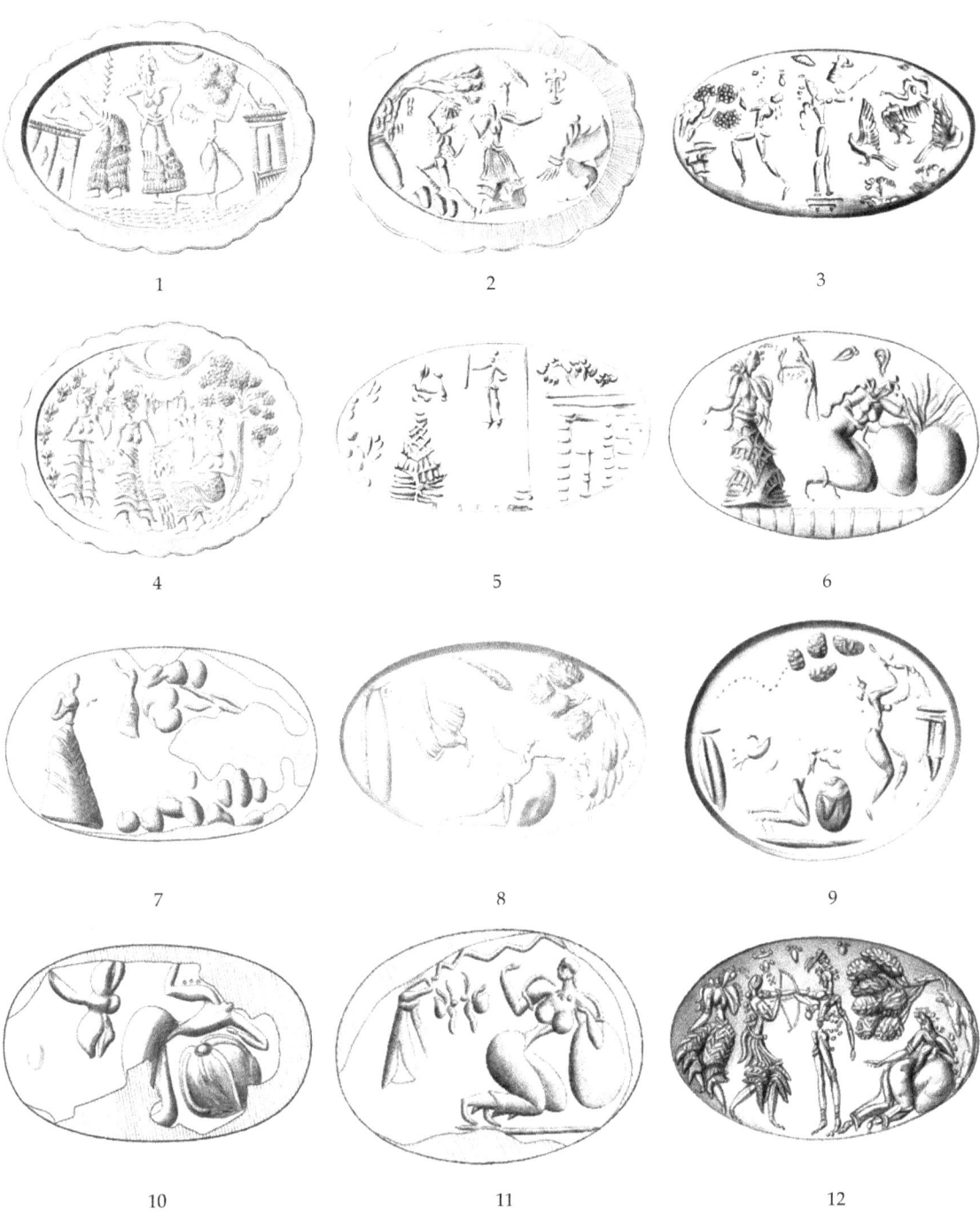

Fig. 31. *(1)* Mycenae gold ring CMSI, 126 *(2)* Vapheio gold ring CMSI, 219 *(3)* Poros gold "New ring" HM 1629 *(4)* Mycenae gold ring CMSI, 17 *(5)* Poros gold "Old ring" AM 1938.1127 *(6)* Vapheio? gold ring AM 1919.56 *(7)* Hagia Triada sealing CMSII6, 6 (HM 522) *(8)* Sellopoulo gold ring HM 1034 *(9)* Kalyvia gold ring CMSII3, 114 *(10)* Zakros sealing CMSII7, 6 *(11)* Hagia Triada sealing CMSII6, 4 *(12)* Berlin gold ring CMSXI, 29

Fig. 31 (cont). *(13) Hagia Triada sealing CMSII6, 2 (HM 523) (14) Chania sealing CMSVSIA, 180 (Chania 1024) (15) Mochlos gold ring CMSII3, 252 (16) Hagia Triada sealing CMSII6, 1(HM 505) (17) Knossos bronze ring CMSII3, 15 (18) Chania sealing CMSVS1A, 176 (Chania 2055) (19) Metropolitan Museum seal CMSXII, 264 (20) Aidonia gold ring CMSVS1B, 114 (21) Makrys Gialos seal CMSVS1A, 55 (22) Zakros sealing CMSII7, 218 (23) Zakros sealing CMSII7, 1 (24) Sealing CMSII6,3*

"Fruits on sacred trees"

Since 1901, the year of the publication of Evans' article "Mycenaean Tree and Pillar Cult", the sanctity of the tree in the Prehistoric Aegean world has been unanimously accepted[196]. For Nilsson this sanctity is unquestionable even if "we cannot always decide with certainty whether the tree is holy on its own account, or as the embodiment of a deity, or simply because it belongs to a sacred grove inhabited by the god or containing his temple"[197].

A common feature of these "sacred trees" is the round objects hanging from their branches. Two of the best known representations of these round objects, the very same ones that Evans used in his article as examples, appear in a gold ring from Mycenae CMS I,126 (fig. 31.1; fig. 32.1) that has been interpreted as a "mourning scene for divine youthful hero"[198], and in a gold ring from Vapheio CMS I,219 (fig. 31.2; fig. 32.2) both dated LHII. Two other typical examples are to be found in the gold "new ring" from Poros (HM 1629)[199] (fig. 31.3; fig. 32.3), and in another famous gold ring from Mycenae CMS I,17 (fig. 31.4; fig. 32.4). These formations are interpreted as "fruits" or "bunches of grapes"[200]. In our view, using strictly shape matching criteria, they could also be pinecones, date clusters or swarms of bees (a swarm of bees resembles a bunch of grapes and for that reason, in antiquity, they had the same name[201]) (fig. 1, fig. 33.1-3).

In order to confine this polysemy we decided that it was necessary to seek a specific agricultural practice that could be used as the sub-context of the primary agricultural context for the analysis of all the other forms appearing together with the round shaped

---

[196]Evans 1901, 99-204, Nilsson 1950, 262, Marinatos 1989, 127-142, Dietrich 1996
[197]Nilsson 1950, 264
[198] Evans 1921-1935, I, 161
[199] Demopoulou et al. 2000
[200] Fruits: Evans 1901, 177; Evans 1921-1935, I, 162. Brunches of grapes: Schliemann 1878, 354; Schuchhardt 1891, 276 (reported by Cook 1965, 515, n 1); Evans 1901, 101; Sakellariou 1966, 32. Others speculate that it is leafage of olive or pine (see for bibliography Cook as above)
[201]Homer (*Il.*2.89) calls the swarm on trees βότρυς which he also uses for bunch of grapes (Eustathius *Ad Iliadem* 1.274.34). Likewise, Varro (*Rust.* 3.16.29), Virgil (*G.* 4.557) and Pliny (*HN* 11.18) compare the swarms with a bunch of grapes. In Byzantine *Geoponica,* the word βοτρυδόν often stands for "*melisostafillo*" ("beegrape"), a term used for the swarm in Greece today. Similarly, honey harvesting, from at least the 2nd c. BCE (Mosch. *Ep. Bion.* 3.35), is called τρύγος (vintage) which also is used for harvesting of vines. The same terminology is used in Byzantium and modern Greece (Kukules 1951, 353; Loukopoulos 1983, 405). Roman writers, too, used the term *mellis vindemia* (Columella *Rust.* 9.15.1) for honey harvesting. Swarms (ancient Greek ζμῆνος, σμῆνος, σμᾶνος, ἐσμός, ἀφεσμός) (Liddell and Scott, *Greek-English Lexicon* s.v.) are formed by the bees when they increase in number and leave their nest in spring (May or beginning of June) (Winston 1987, 182) and as a lot fly to the nearest branch of a tree where they shape a cluster-like pattern with their queen in the center. Bee-scouts are then sent out to find a suitable location for the new hive. The may remain on the branch for a few hours or a few days. The swarm then flies away to its permanent nest which is usually some tree or rock cavity. The image of a swarm attached to a brunch of a laurel, is given by Livy (*Epit.* 21.46.2) and, most vividly, by Virgil (*Aen.* 7.64-66): "*A swarm of bees that cut the liquid sky, (Unknown from whence they took their airy flight), Upon the topmost branch in clouds alight, These, with their clasping feet, together clung, And a long cluster from the laurel hung.*" Dryden J. 1697. ed. and trans. *Virgil Aeneid*

objects on the trees. In other words we tested our primary identification of the round shaped objects with the above mentioned alternatives by examining their interrelations with all the other forms that appear in every signet ring that represent round shaped objects. We firstly excluded the interpretations of big leaves and pinecones since there are no agriculture practices for such products.

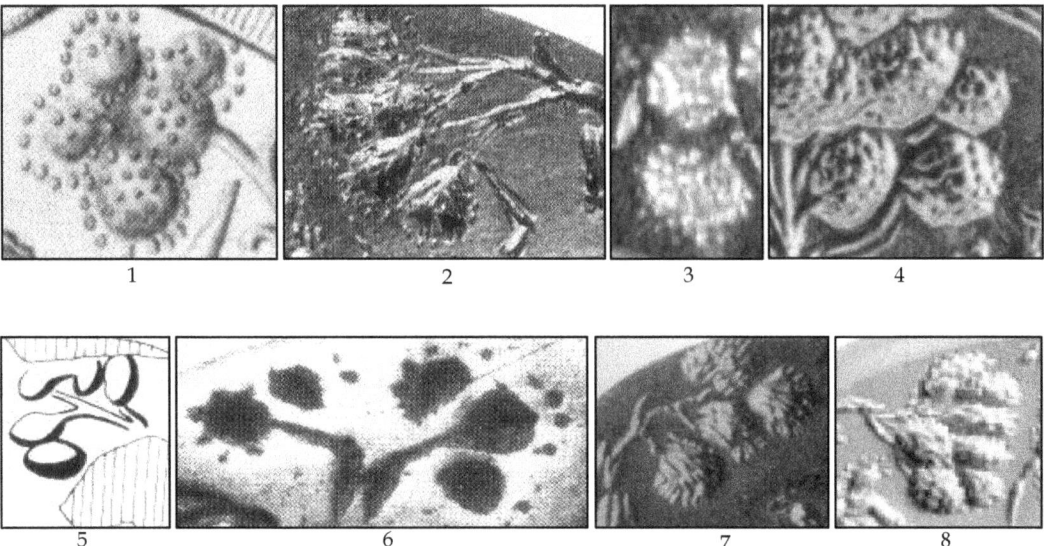

Fig. 32. Swarms of bees depicted as rounded objects on trees (1) Mycenae CMSI, 126 (2) Vapheio CMSI, 219 (3) Poros HM 1629(4) Mycenae CMSI, 17 (5) Hagia Triada CMSII6, 6 (6) Mochlos CMSII3, 252 (7) Sellopoulo HM 1034 (8) Berlin CMSXI, 29

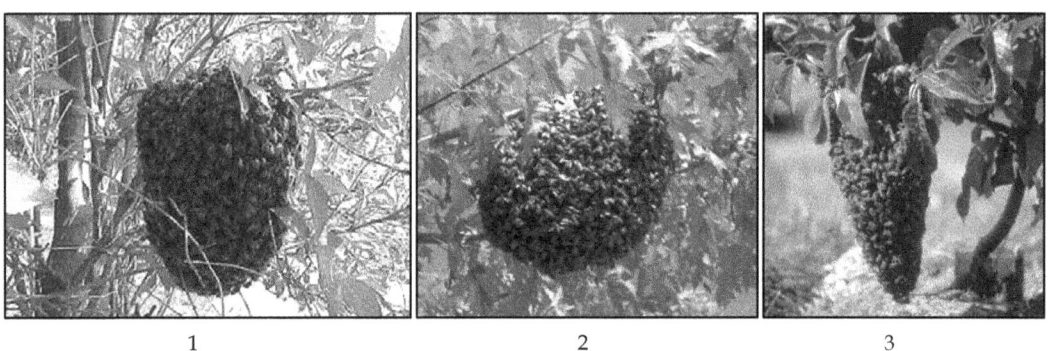

Fig. 33. Swarms of Apis mellifera on trees (photos (1): http://www.beemaster.com (2): http://www.oznet.ksu.edu (3): http://www.permaculture.co.uk)

A helpful thread for further narrowing the possibilities is the fact that these rings imply a degree of wealth in terms of the materials used and the high quality of craftsmanship evident in their construction. Certainly, they must be seen as an exclusive class of objects associated with the most privileged stratum of the Aegean Bronze Age society. It is obvious that such privileges emanated from the high value of the product that the owners of the rings traded or controlled. That led us to exclude the date cluster scenario; even though the date palm probably existed in Minoan times, it is hardly possible that its production attained significant proportions. Fruit-growing of course and viniculture were important for Minoans and Mycenaeans but so was apiculture, and maybe even more important, as we have shown in the first part of our monograph. We were able to exclude the fruit-growing and viniculture contexts since we were not able to obtain an intelligible meaning when we tried to fit in the other forms depicted in the group of scenes with trees carrying round objects on their branches. That leaves us with the apiculture context and the swarms of bees. How do the rest of the forms appearing along with the round objects on the trees fit in an apiculture context?

"Festoons and pearls"

In the CMS I,126 gold ring (fig. 31.1), dots (usually called "festoons" or "pearls"[202]) are depicted around the round shaped objects (fig. 32.1). The same dots exist on the upper arms of the central female figure as well as on the back of the left female figure and on their dresses (fig. 34.1-2). The same distribution of dots is seen in Vapheio CMS I,219 (fig. 31.2; fig 34.3). We believe, in conjunction with the argument that the round shapes on the trees are swarms, that these dots actually represent bees.

Fig. 34. Bees depicted as dots (1, 2) Mycenae CMSI, 126 (3) Vapheio CMSI, 219

The technique to represent bees as dots is, from a technical point of view, practical and easy and for this reason quite common; it is encountered in Mesolithic art as well as in modern one (fig. 35)[203].

Fig. 35. Bees depicted as dots (1) Capture of a swarm of bees from a tree and a nearby beehive (Hachimitsu-ichiran 1872, Japan) (2) St. Sossima and St. Savati capturing a swarm from a tree (Armbuster 1928).

---

[202] Evans 1921-1935, I, 161 n4 and I, 494; Nilsson 1950, 275
[203] In Mesolithic and later rock paintings in India (Crane 2000a, 37, fig 10.2b, fig. 10.2c, fig. 19.2c) and in Africa (Crane 2000a, 50, fig. 8.1c)

Another motif from CMSI,126 (fig. 31.1) and CMSI,17 (fig. 31.4) are the thin lines in the form of an arch (fig. 36.1-2). These lines have strong similarities with the catenaries curves representing an *Apis mellifera* nest in caves, seen from the front, as can be seen on rock paintings from Africa[204] (fig. 36.3-5). In front of a cave is believed to take place the scene of CMSI,17 from Mycenae[205].

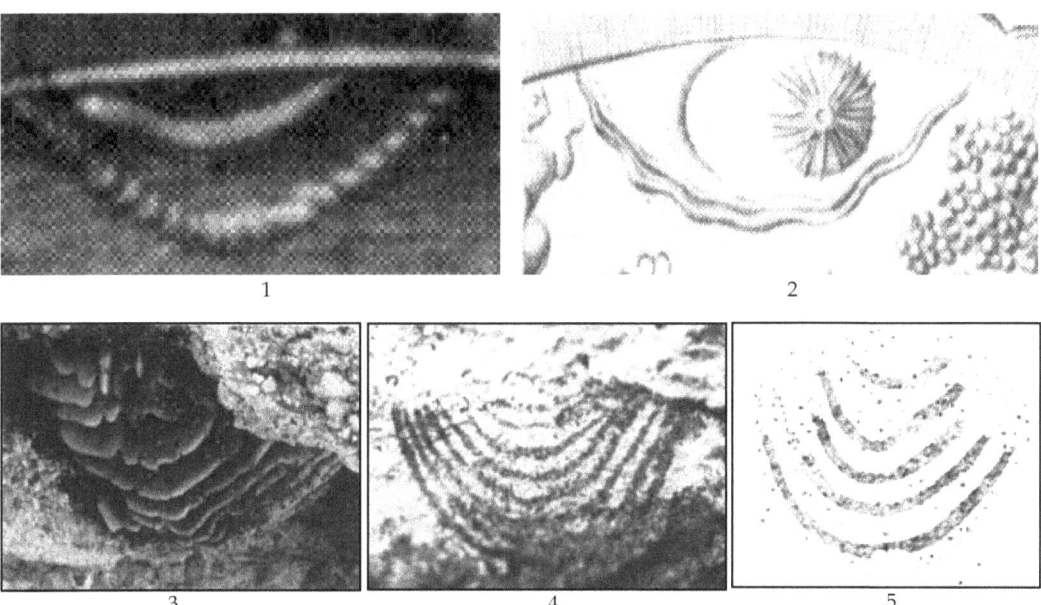

Fig. 36*. Catenaries curves and wild bee nests in caves. (**1**) CMSI, 126 (**2**) CMSI, 17 (**3**) Actual nest of Apis mellifera at the entrance of a cave (photo Loper G.M. from Crane 2000a, 49, fig. 8.1b). (**4**) Catenary curves representing an Apis mellifera nest from a rock painting in S. Africa (photo Guy R. from Crane 2000a, 49, fig. 8.1a). (**5**) Rock painting from S. Africa Representing an Apis mellifera nest (Photo R. Guy in Pager 1971, from Crane 2000a, 50, fig.8.1c)*

---

[204] Crane 2000a, 49-50, fig. 8.1a, 8.1b, 8.1c
[205] Danielidou 1998, 149

"Ritual bending of the sacred tree"

In the aforementioned rings, human figures stand next to the so-called "sacred" tree, bending one of its round object bearing branches (fig. 37.1-4).

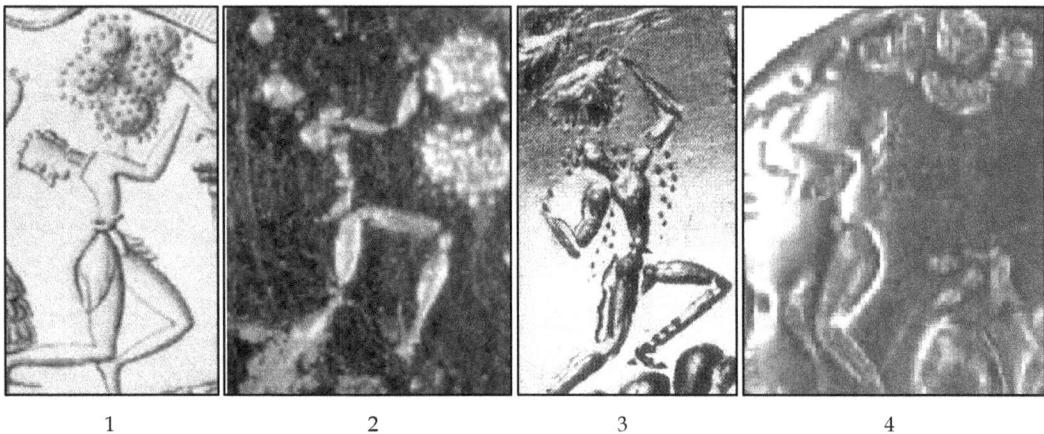

Fig. 37. Swarm capture (1) Mycenae CMSI, 126 (2) Poros HM 1629 (3) Vapheio CMSI, 219 (4) Kalyvia CMSII3, 114

This motion has been described as "ritual bending of the sacred tree" or as "bringing down the sacred tree"[206]. But, if the cluster-like objects on the trees are indeed swarms, this so called "ritual action" would simply be the capture of swarms from the trees, an action similarly depicted in later sources (fig. 35.1-2; fig. 38.1-4). Capturing a swarm during the swarming period is vital for beekeepers[207] and various ways have been invented since antiquity in order to achieve this. To capture a swarm that has already settled on a branch it would suffice to shake or cut the branch causing the swarm to fall in a basket placed underneath[208]. A swarm is actually quite docile, more so than when it is hived. Reportedly, a beekeeper may capture a swarm by reaching into the mass of bees with his hand and picking out the queen bee; the swarm will follow wherever he takes her. In Rhodes, traditional apiarists used to attract swarms in small baskets ("asmarolo") rubbed with leaves of sage[209]. If the swarm was settled on a high branch, the only way to capture it was to place an empty hive under the tree so that bees would get into it by themselves[210] or, as they used to do in Florina (Northern Greece), to hang a hamper in the tree in order for the swarm to enter it[211]. In classical Greece (Pl. *Laws* 843d), as in Roman and Byzantine periods special laws existed that regulated the right of capturing a swarm from a tree[212].

---

[206] Glotz 1923, 286; Nilsson 1950, 43, 275; Evans 1901, 177
[207] See Part 1 "Beekeepers and beekeeping practices"
[208] Nikolaidis 2000, 38, Kostakis 1963, 388
[209] Vrontis 1939, Oikonomidis 1965-1966, 632
[210] Nikolaidis 2000, 116-117
[211] Anagnostopoulos 2000, 307
[212] For the Byzantine period, see Kukules 1951, 355; Liakos 1999b. For the Roman period, see Crane 2000a, 208-9; Frier 1982/1983. The Roman law on swarm ownership formed the basis of rulings for many countries (whose law belongs to the Romano-Germanic family) such as Austria, Bulgaria, Czechoslovakia, Sweden, Finland, France, Germany, Italy, Portugal and

According to St. John Chrysostom (*In epistulam ad Ephesios* PG 62.105) apiarists in Byzantium employed a method for capturing the swarm resembling the one mentioned above, still in use in modern Florina. A report in *Mythologikon Syntipa* (28.10.14) mentioned a hunter who captured a swarm with a basket (εὕρε μελίσσιον μετά κοφίνου). There exist testimonies for similar practice in the Roman period (Columella *Rust.* 9.8.11) and in Hellenistic times as indicated by the phrase in Strabo (2.1.14): ἐν τοῖς δένδρεσι σμηνουργεῖσθαι, considering that σμηνουργοί in antiquity designated those who captured swarms (Ael. *NA* 5.13.14).

Fig. 38. Capturing a swarm of bees from a tree (1) Byzantine depiction (2) Alphandery 1911 (3) Debeauvoys 1846 (4) Modern scene of capture of a swarm from a tree, in Greece (photo T. Bikos from Bikos 1996, photo 11)

Switzerland. Similar laws existed in Argentina, Chile and Brazil (Crane 2000a, 209). Similar laws exist in the *Talmud* (*Baba Qamma* 10.2; *Shabbath* 43b) as in many regions of Greece up to recently (Petropoulos 1957, 190, n 1)

"Flying gods"

In further developing the apicultural context we consider certain figures appearing in the scenes. In the CMS I,17 there is a male figure shown full face "hovering in the air" and carrying an eight shaped shield (fig. 31.4; fig. 39.1), identified as a "god of the Minoan pantheon"[213]. A similar figure, shown in profile, can be seen in the Ashmolean Museum gold ring Nr. 1938.1127 ("old Poros" ring) (fig. 31.5, fig. 39.2) and in the Ashmolean Museum gold ring Nr. 1919.56 (fig. 31.6; fig. 39.3). We believe that we are presented here with three cases of the so called cavalier perspective (a technique of perspective, often used in Minoan iconography, where figures that are further up are meant to be further away[214]) combined with an intuitive perspective of size diminishing[215]. We estimate that these figures represent Kouretes beekeepers striking their bronze shields to make the swarms settle, an ancient apiculture practice called tanging that prevailed until recently among beekeepers in Greece and in many other parts of the world[216].

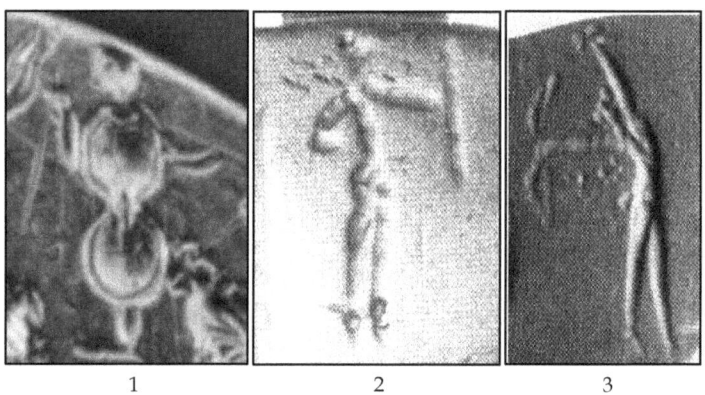

Fig. 39. Tanging (1) Mycenae CMSI, 17 (2) AM 1938.1127 (3) AM 1919.56

---

[213] Nilsson 1950, 347
[214] For the cavalier perspective see Walberg 1986
[215] This is also the opinion of Mylonas (1966, 149-50, 157, 169-71)
[216] For the Kouretes and tanging see Part 1 "Beekeepers and beekeeping practices"

"Great Goddess"

The same as above cavalier/diminishing size perspective we recognize in the way that the two female figures (centre and left) are represented in the gold ring from Poros (fig. 31.3; fig. 40.1-2). An intuitive perspective or simply small young girls can also be recognized in the two small sized female figures of CMSI,17 (fig. 31.4; fig. 40.3-4) occurring along with three other women (fig. 40.5-6). Female figures appear also in Vapheio CMSI,219 ring (fig. 40.7), in Mycenae CMSI,216 ring (fig. 40.8-9), in AM 1938.1127 (fig. 40.10) and AM 1919.56 gold ring (fig. 40.11). On a sealing from Hagia Triada CMSII6, 6 we see a central female figure, a smaller one further away (cavalier perspective) and swarms on the tree (fig. 31.7; fig. 32.7; fig. 40.12).

It is known that, traditionally, both men and women were engaged in beekeeping and according to literary sources, the same stands for antiquity[217]. Could then the women of the rings be beekeepers too thus explaining why some of them are surrounded by bees?

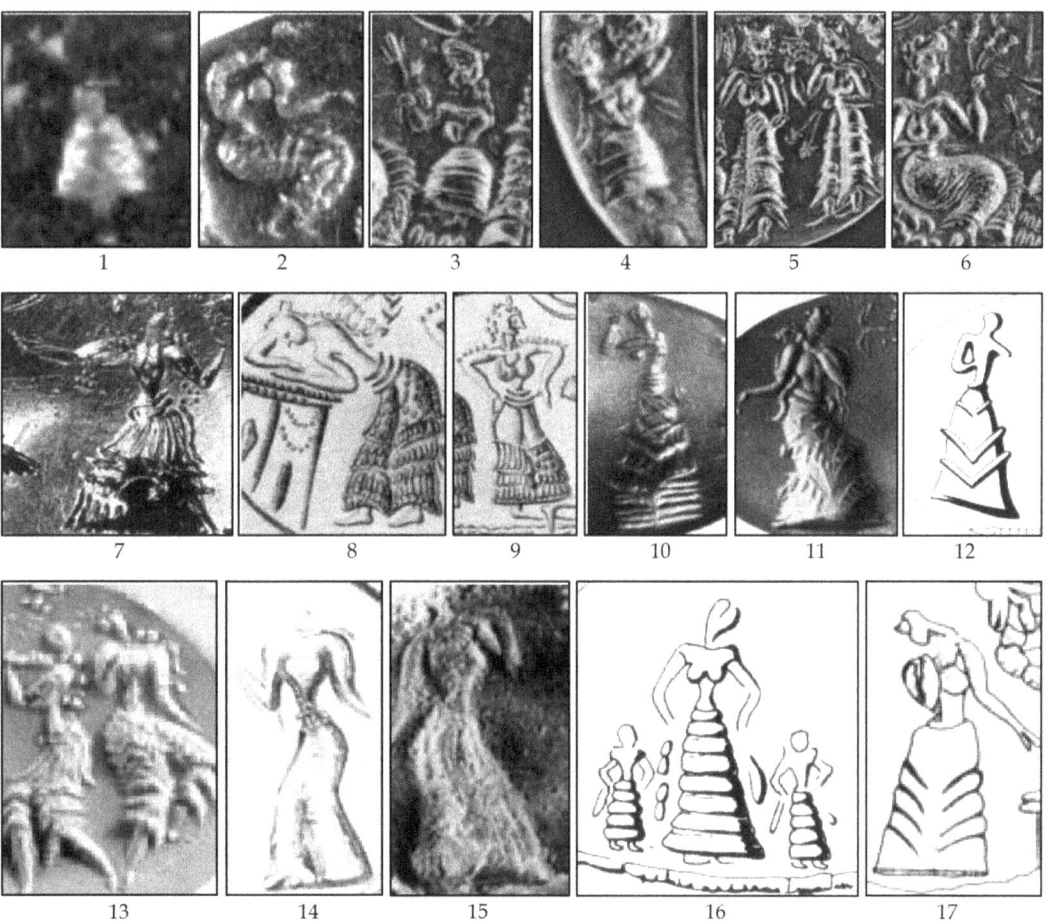

Fig 40. Women beekeepers *(1, 2)* Poros HM 1629 *(3, 4, 5, 6)* Mycenae CMSI, 17 *(7)* Vapheio CMSI, 219 *(8, 9)* Mycenae CMSI, 126 *(10)* AM 1938.1127 *(11)* AM 1919.56 *(12)* Hagia Triada CMSII6, 6 *(13)* Berlin CMSXI, 29 *(14)* Chania 1024 *(15)* Hagia Triada CMSII6, 2 *(16)* Hagia Triada CMSII6, 1 *(17)* Chania 2055

---

[217] See Part 1 "Beekeepers and beekeeping practices"

A common feature of these female figures (or "Great Goddesses" as they are called) is the well known Minoan bell like flounced skirt, also called "sacral dress"[218] generally ascribed to "priestesses", due to the religious context given to the scenes by Evans. It is certain that a dress in prehistory was more than a simple item of clothing. Renfrew recognized in the dresses of prehistoric Aegean frescoes "a central role in some of the actions depicted"[219] and in Classical Crete a distinctive costume denoted a special status: it was a new garment that signalled the assumption of the role of someone as warrior, hunter, or citizen (Ephorus fr. 149; Arist. *Pol.* 1271b, 1329b; Strabo 10.3.11). Are there any elements that could lead us to match this particular Minoan apparel with a woman beekeeper? The ambiguity that results from the absence of colours and the miniature nature of the scenes poses considerable difficulties for a safe answer. There are however two intriguing details on a colourful and bigger example: the dress of the faience statuette "Goddess with the snakes" (MMIIIb) (fig. 41). This dress consists of seven tiers decorated in alternating blocks in tones of yellow and black[220]. This is not an isolated example; the contemporary gold pin head from Shaft Grave III from Mycenae represents an exact parallel of this seven tier skirt. Prototypes of the flounced skirt appear in the ancient Near East and the robes of the "presentation goddess" on the Investiture Painting from Mari (c. 1750 BCE) have tiers painted in colours similar to those of the Minoan statuette[221]. Is it just a remarkable coincidence that the bee abdomen has the same number (seven) of visible segments (as the bees of the Mallia pendant and in the CMS II5,314) with alternating yellow and black colours? Does, then, this particular clothing allow the identification of a clear role: a woman beekeeper, a "nymph", the female counterpart of Kouretes[222] who we identified on the same rings as the performers of tanging? Of course in the above rings other types of flounced dresses appear that do not have, or it is not always evident if they do have seven tiers. We could assume that this theme variety indicates different hierarchical posts in a beekeeping association or that the engraver knew that no more than a few initiated observers were to read the meaning of scenes and thus no obstinacy to details was needed[223].

Fig. 41

The Snake Goddess (photo Hrakleion Museum)

---

[218] Nilsson 1950, 155-156

[219] Renfrew 1999, 712

[220] Evans (1921-1936, I, 503) identifies the colours as "dark orange" and "purplish-brown". Jones (2001) identifies the colours as "dark ochre" and "purplish-blue" but she admits that the colours are degraded.

[221] Jones 2001

[222] See Part 1 "Beekeepers and beekeeping practices"

[223] Morgan 1989, 146

"Baetyls and bird epiphany of deities"

The second to the right female figure of the AM 1919.56 ring is leaning on two enigmatical round objects from which several "lines" project upwards (fig. 42.1). The same form – a figure leaning on a round object – is encountered in the gold ring from Sellopoulo (tomb 4) of Knossos HM 1034 (fig. 31.8; fig. 42.2), in the gold Kalyvia ring CMS II3,114[224] (fig. 31.9; fig. 42.3), in the Zakros sealing CMS II7,6 (fig. 31.10; fig. 42.4), in the Hagia Triada sealing CMS II6,4 (fig. 31.11; fig. 42.5) and in the Berlin museum ring CMS XI,29 (fig. 31.12; fig. 42.6). Similar round objects can be seen in the sealing from Hagia Triada CMSII6, 2 (fig. 31.13; fig. 42.7), the sealing CMSVS1A, 180 from Chania Nr.1024 (fig. 31.14; fig. 42.8), the the ring CMSII3,252 from Mochlos (fig. 31.15, fig. 42.9) - in a cavalier/diminishing size perspective - and maybe in the Hagia Triada sealing CMSII6, 1 (fig. 31.16; fig. 42.10).

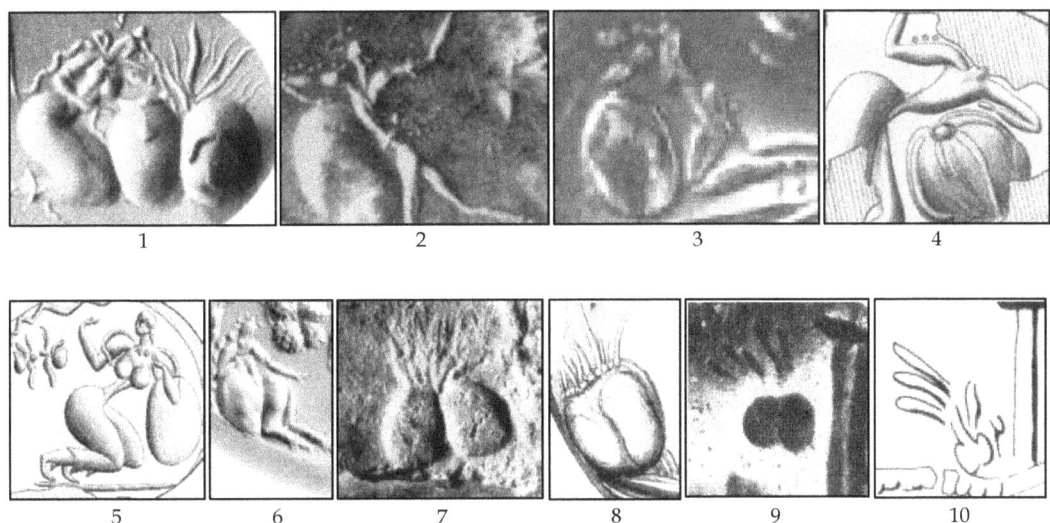

***Fig. 42**. Bee sacs (**1**) AM 1938.1127 (**2**) Sellopoulo HM 1034 (**3**) Kalyvia CMSII3, 114 (**4**) Zakros CMSII7,6 (**5**) HagiaTriada CMSII6,4 (**6**) Berlin CMSXI,29 (**7**) Hagia Triada CMSII6, 2 (**8**) Chania 1024 (**9**) Mochlos CMSII3, 252 (**10**) Hagia Triada CMSII6, 1*

These objects are known as "bilobar objects that are connected with the lamentation of death and the vegetation circle of life and death", "jars", "pithos burial", and the plant Sea Squill or "stone baetyls"[225]. Before proceeding further to the analysis of these objects we must underline the presence of other forms that either we have already interpreted in a beekeeping context or can be easily identified as such. In both the Asmolean 1919.56 and Berlin rings (fig. 31.6; 31.12) we note the presence of three female figures surrounded by dot-bees wearing the characteristic dress which we identified above with women beekeepers' (fig. 40.13; fig. 42.1).

---

[224] Platon 1984 ; Nilsson 1950, 268, fig.133
[225] Papapostolou 1977, 82; Persson 1942, 35; Cain 2001, 37, fig.13. For a complete review of the theories about these round objects see Warren 1984, 17-24 and Papatsaroucha 2005,177-183

The female figures on the Chania 1024, and Hagia Triada CMSII6,2 and CMSII6,1 sealings (fig. 40.14-16) could also represent women beekeepers. In the Hagia Triada CMSII6,1 the female figures are surrounded by dot-bees and there is also a tree with swarms. In the Berlin museum ring the round object is placed below a tree with swarms (fig. 32.8). In the Kalyvia ring there is a male figure that captures swarms from a tree (fig. 37.4). In the sealing from Hagia Triada CMSII6,4 we can actually see two bees flying[226] (fig. 43) and in the Mochlos ring we can see a tree with swarms and dot-bees around them (fig. 32.7). A new form appearing in the Sellopoulo, Vapheio, Aidonia and Kalyvia rings are the "floating spikes"[227] (fig. 44.1-3) which we could easily identify with the formations of airborne swarms (fig. 44.4-5).

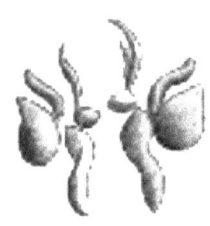

Fig. 43
Detail from Hagia Triada CMSII6,4

In the Sellopoulo ring the round object is next to a tree with swarms (fig. 32.7) and the male figure that embraces the round object is surrounded by dot-bees (fig. 42.2). It appears as if he was trying to protect the round object from a bird charging with speed towards it (fig. 45.1). A similar scene of a charging bird we see in the sealing CMS II7,6 from Zakros (fig. 31.10, fig. 45.2). A bird appears also in the scene of the Kalyvia ring (fig. 45.3) and two other birds appear as if they were driven away by the central male figure in the new Poros ring (fig. 45.4-5). This bird presence on the rings is considered as "bird epiphany of a deity"[228]. However, the characteristic big tail of the birds in the Sellopoulo ring and in the Zakros sealing is also the main characteristic of the long-tailed tit (αἰγίθαλος) (fig. 45.6) a bird known to Aristotle (*Hist. An.* 592b) as a bird with big tail (οὐραῖον μακρόν ἔχων) and as a bird preying on bees - a basic enemy of ancient and modern apiarists (Arist. *Hist. an.* 626a; Ael. *NA* 1.58; *Geoponica* 15.2.18)[229]. Other birds on the above seals if not long-tailed tits too, could represent other known (Arist. *Hist. an.* 626a; *Geoponica* 15.2.18) bird bee-enemies, such as merops (*Merops apiaster*)[230] (fig. 45.7). That is why the apiarists on the rings are trying to drive away the birds, to protect their bees, like the watchman in Virgil's *Georgics* (4.110) who keeps the beehives from birds and thieves.

But why are they protecting the round objects? There is only one logical explanation: the long-tailed tit (or merops) covets swarms just been captured from the tree and placed in the round objects. So these objects must be some kind of beekeeping items, and in particularly bee sacs intended as receptacles of captured swarms. In antiquity such an item was the cow's stomach. Its use in apiculture is not an oddity. Mago and Celsus report that bees can "*come out*" of the stomach of an ox (Columella *Rust.* 9.14.6; Plin. *HN* 11.23) and Varro (*Rust.* 3.16.15) draws our attention to the fact that the word for the beehive in

---

[226] They are usually characterized as a butterfly (Evans 1921-1935, III, 148; Kyriakides 2005, 147)
[227] Kyriakidis 2005, 140, fig.3
[228] Nilsson 1950, 333
[229] Oikonomidis 1965-1966, 627
[230] Foster (1995, 415-18) identifies the same birds as swallows. Even if her view is right, this does not affect our argument because swallows as merops are both preying on bees. Merops can confidently be identified in Egyptian art and hieroglyphs (Houlihan 1986, fig. 166).

Latin (*alvus*[231]) is the same as the Latin word for stomach. Varro also reports that certain beehives are manufactured in a manner imitating the bipolar form of a stomach[232]. Apiarian books of 14th c. advice the use of an ox stomach as beehive[233].

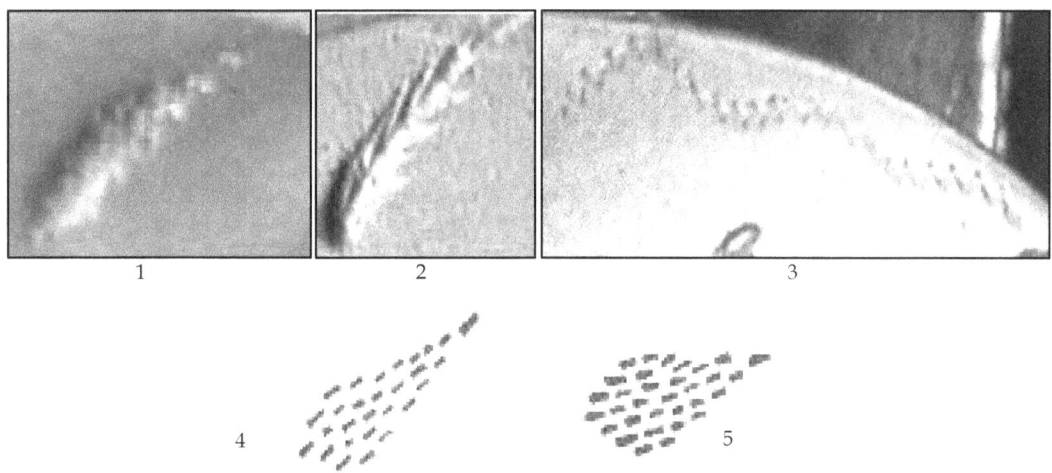

Fig. 44. Flying swarms (1) Sellopoulo HM 1034 (2) Vapheio CMSI, 219 (3) Aidonia ring (4, 5) Flying swarms (Hannemann 1850. "Die Befruchtung der Konigin". Bienen Zeitung 6:12-3; 19-20)

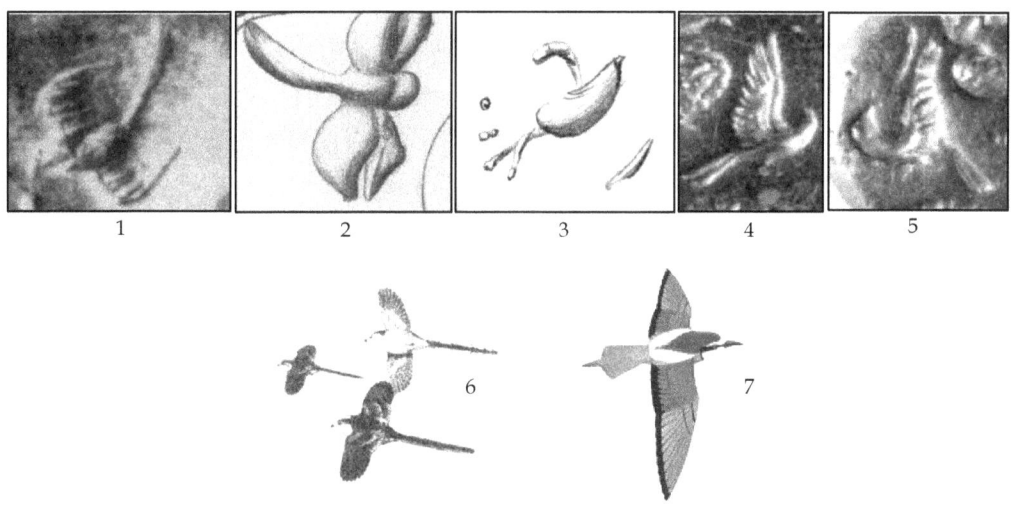

Fig. 45. Birds preying on bees (1) Sellopoulo HM 1034 (2) Zakros CMSII7, 6 (3) Kalyvia CMSII3, 114 (4, 5) Poros HM 1629 (6) Long-tailed tit (7) Merops apiaster

---

[231] "*alvus*": Columella *Rust.* 9.2.1, 9.6.14, 7.15; Plin. *HN* 11.22.23.69, 21.80.82. "*alveus*": Tib 2.1.49; Columella *Rust* 9.3.1.
[232] According to the interpretation of Hooper W. D. 1960 *Marcus Terentius Varro on Agriculture*. Revised by Ash H. B. Harvard University Press, 508, n 1
[233] Ransome 1937, 153 for references. Traditional beekeepers in the Aegean used the *asmorolo*, a sac made of tule (Korre-Zografou 2008, 201)

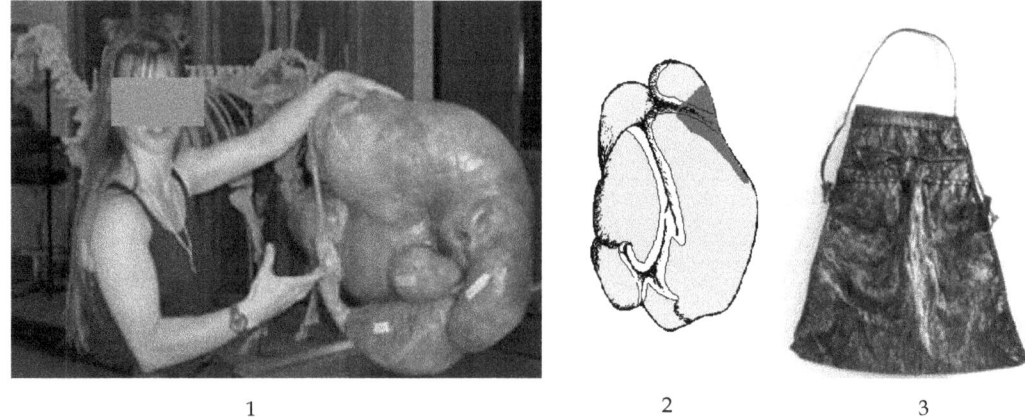

Fig. 46. The stomach of the cow (1) (photo http://www.its.caltech.edu) and (2) (photo: http://137.222.110.150/calnet/abdpel9/page17.htm) (3) Bee-sac made of a cow stomach (photo J. Corner from Crane 2000a, 58 fig. 8.5b)

The stomach of the ox comprises two large compartments (and two less marked and much smaller ones) which recall the bilobar form of the object on the seals. Moreover the stomach of an ox has a capacity of 150-250lt and a length of about 70cm[234], dimensions that match the dimensions of the round objects on the above seals (fig. 46.1-2). Thus, the round objects could be sacs made from cow stomach, for putting in the captured swarms[235] (fig. 46.3). But sacs have usually straps for carrying them and this is precisely what we recognize in the projecting lines of the round objects. Indeed, the female figure of Hagia Triada sealing CMS II6,4 (fig. 42.5) seems as she has passed around her shoulders two straps of the sac and she is ready to carry the sac away. Two pairs of straps can also be seen in the Zakros CMSII7,6 sealing (fig. 42.4) which appear to be tied, closing the sac with a purse-string mechanism.

---

[234] Barone 1984, 335-375

[235] The Akkadian and Sumerian word for stomach is *"karsu"* (Black 2000) and perhaps we should recognize in this word the Hittite *"kursa"*, the sac that brings the bee to the Goddess Inara (KUB XIII 59 iii, 5-13). The *"kursa"* was made of skin of animals but also of other materials as buckram, timber, straw but also lapis lazuli (Morris 2001, 144). Perhaps other materials replaced the initial use of ox's stomach when need imposed it (see also the Akkadian word *"kursallum"* meaning the basket, in Black 2000 - for resembling types of the word in the Semitic languages see Militarev 2000, 136). The word *"kyrserida"* which according to Hesychius (s.v. κυρσερίδες) means the beehive could be also etymologised, in our opinion, from the Hittite *"kursa"* or the Akkadian *"karsu"*. The word κύστεροι according to a gloss from Hesychius (s.v.) means also the beehives. The word obviously emanates from the word bladder (κύστις) that means the sac made of animal tissue.

"Altars"

Another prominent motif of the rings in question constitutes the so called "altar", the rectangular construction that usually stands below a "sacred tree". Here too it is possible to recognize certain important beekeeping paraphernalia: beehives. In the first part of this monograph we have shown that Minoan beekeepers used ceramic (horizontal and upright) beehives and probably woven wicker beehives too. We propose that we should recognize as horizontal ceramic beehives firstly the object in Vapheio ring (with projecting rims?) (fig. 47.1, fig. 4) standing (ready to accept the swarm of bees) below a tree[236], and secondly the object in the bronze ring CMSII3,15 from Knossos (fig. 31.17; fig. 47.2) above which there are some dot-bees leaving/entering the vessel.

Another type of "primitive" beehive is the stone beehive. Its existence in antiquity is implied by Homer (*Od.* 13. 106-7) and by *Leiden demotic papyrus* (3rd c.)[237]. The simplest type of stone beehive consists of four stone plates forming a rectangle, and two movable bark plates serving as covers to the sides[238]. William Leake reports stone beehives in 1805 in Mani and in 1790, Rocca reported their existence in Syros[239].

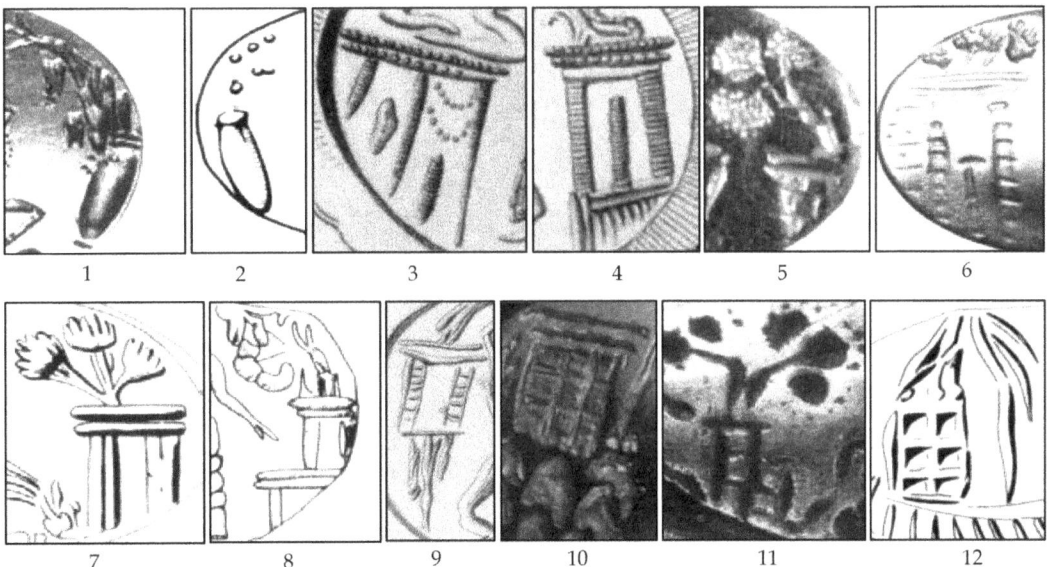

Fig. 47. Beehives (1) Vapheio CMSI, 219 (2) Knossos CMSII3, 15 (3, 4) Mycenae CMSI, 126 (5) Poros HM 1629 (6) AM 1938. 1127 (7) Hagia Triada CMSII6, 1 (8) Chania CMSVS1A, 176 (Nr. 2055) (9) CMSXII, 264 (10) Aidonia ring (11) Mochlos CMSII3, 252 (12) Makrys Gialos

---

[236] Tsountas (Αρχ. Εφ. 1890, 170) characterizes this object as a "big container", Evans (1901, 176, and 1921-1935, 432) as a "stone column" and Persson 1942, 36 as "burial jar". Nilsson 1950, 275 characterizes it as "unknown object"
[237] Forbes 1966, 86
[238] Vrontis 1939, 195, fig. 1
[239] Leake 1830; Rocca 1790

Stone hives in wall recesses, called "melissothyrides" or "smodohia", existed in the Cycladic islands such as Paros, Andros, Tenos, Naxos and Sikinos but also in Leukas[240] and in mainland Greece, in Mycenae. Stone beehives can still be seen in Greece (fig. 48.1-3)[241]. We think that this kind of beehive, open in the front side, is depicted in the ring CMS I,126 (fig. 31.1; fig. 47.3) beneath a woman beekeeper who bends over it. This object is usually recognized as an "altar" or "shrine"[242].

Fig. 48. *(1) Stone beehive from Kythera (Protopsaltis 2000, 290 fig. 4).(2) Stone beehives in wall recesses in Tenos (photo T. Bikos from Bikos 1997, 487, fig. 4) and in (3) Kythera (Rammou 2000, 426).*

Fig. 49. *A Byzantine representation of swarm capture. At the base of the second from the left tree, a horizontal beehive is placed*

Another stone beehive can be seen in the opposite extremity of the ring, below the tree with swarms (fig. 47.4). Similar constructions can be seen in the new Poros ring (fig. 31.3; fig. 47.5), the AM 1938. 1127 ring (fig. 31.5; fig. 47.6), the Hagia Triada CMSII6, 1 (fig. 31.16; fig. 47.7), the Chania 2055 ring (fig. 31.18; fig. 47.8) – next to a woman beekeeper - and in the CMSXII,264 with a male figure trying to capture a swarm (fig. 31.19; fig. 47.9). The rectangle structure near a tree and over (or maybe between) rocks depicted on a ring from the Aidonia woman's grave 7 (fig. 31.20; fig. 47.10) could represent an open stone hive in a wall recess while the three women with flounced skirts standing in front of the hive could represent beekeepers. It is interesting to note that these "altars" are almost always placed near or below "sacred trees". Placing the beehives by the trees, as depicted on a Byzantine swarm capture scene (fig. 49), is recommended by Aristotle (*Hist. an.* 627b, 16), by Roman writers (Verg. *G.* 4.20-24; Plin. *HN* 21.41) and by modern apiarists[243] because trees attract swarms on their branches and

---

[240] Phiorou 2005, 51
[241] Bikos 1997c, 76, fig. 6; Bikos 1997b, 487, fig. 4,5,6; Kukules 1951, 351; Crane 2000a, 316, fig. 32.2g. For the same practice in Britain, Ireland and France see Crane 2000c
[242] Evans 1901, 177; Evans 1921-1935, I, 161; Nilsson 1950, 256
[243] Oikonomidis 1965-1966

because the shade they provide protects the bees. Both ancient and traditional apiarists preferred to plant dwarf trees or tree-like bushes near their hives[244] to facilitate the capture of a swarm settled on its branch. The tree-like bush *Medicago arborea*[245] which flowers in late spring was especially recommended (Arist. *Hist. an* 627b. 17; Dioscorides *De materia medica* 4.112.1, Columella *Rust.* 9.14.5; *Geoponica* 15.2.6).

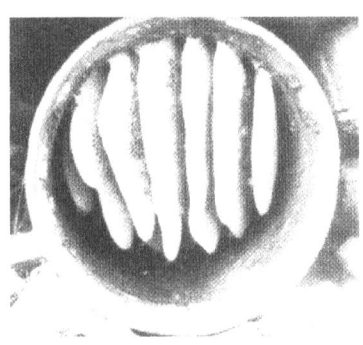

Fig. 50 Sword type of comb in a traditional horizontal beehive (Tenos) (photo N.Karagiorgis)

To the left of CMSI,126, inside the hive open at the side, several dot-bees can be seen (fig. 47.3) and given the detail and conformity that characterizes the Minoan and Mycenaean rings the identification of combs contained inside the stone beehives is quite possible (fig. 47.3-4). These objects have been previously considered as "small baetylic pillars"[246]. Apiarists whom we have consulted recognized in the object the so-called "glossa" (tongue), which is the initial part of the comb raised vertically in the hive before it begins to unfold to the adjacent walls. The fact that the seal represents a spring scene (swarms on the trees) is an additional element supporting the "glossa" theory, since the combs in the hive are shown early in their formation. Alternatively, the object could also represent a specific type of comb, the long linear comb. As Columella (*Rust.* 9.15.6-8) says, the form of the comb depends on the species of the bees and on whether the form of the hive is square, round or elongated. In the traditional horizontal beehives, the bees built two types of combs: round in the form of a disk[247], vertical to the oblong axis of the hive, and long linear combs, parallel to the oblong axis of the beehive, in the form of a sword[248] (fig. 50). For this reason, combs of the latter type were called in Cyprus, "spathakies"[249] (swords) and in Dodecanese and in Cyclades, "mahairia" (knives) or "spathates" or "mahairates" (sword-like or knife-like)[250]. Similarly, in Jordan and in Iran they call these combs "swords", and in Turkey, "arrows"[251].

But a basic disadvantage of stone beehives was the difficulty of transporting them whenever necessary as in the case of migratory beekeeping by sea, the existence of which in prehistoric Crete we have claimed in the first part of the monograph. Hives most suitable for migration were sturdy and light by weight such as those made of woven wicker or wooden

---

[244] Lombard 1812, 73. In Egypt the traditional apiarist prefer to plant cotton (Kueny 1950, 89)
[245] For the identification of *Medicago arborea* with the ancient κύτισσος see Gennadios 1914, 651. Κύτισσος appears in linear B tablets (see Chadwick 1976, 148)
[246] Evans 1901, 177; Evans 1921-1935, I, 161; Nilsson 1950, 257
[247] Disk-shaped combs can be seen on the wall painting in Luxor (fig. 11) portraying an offering of honeycombs on which two bees are seated
[248] Oikonomidis 1965-1966; Crane 2000a, 179-180
[249] Cyprus : Konstantinidis 2000, 325
[250] Dodecanese : Nouaros 1934, 82; Vrontis 1939, 202. Cyclades (Naxos): Oikonomidis 1965-1966. In Naxos they call also "*spathate*" the upright comb in the hive (Oikonomidis 1965-1966, 626) just like "*glossa*"
[251] Crane 2000a, 179, tb. 21.4B

boards (fig. 51.1-3). One ring from Mochlos (fig. 31.15) and one seal from Makrys Gialos (fig. 31.21) may depict the sea transport of wooden beehives.

Fig. 51. (1, 2) Traditional board beehives from Florina, Greece (photo Bikos 1997a, fig. 10, (3) Log hive with bars and board hives with frames, Germany (Forderkreis der Naturwisswenschaftlichen Museum Berlins, from Crane 2000a, 431, fig. 41.3b)

Both scenes have been previously interpreted as an "epiphany" of the "Great Goddess"[252] but there is good reason to believe that they depict beekeeping practice instead. On the ring from Mochlos, there is a boat, a rectangular structure that we identify as a wooden beehive, with a tree behind the hive and swarms on its branches with characteristic dot-bees around them. A woman beekeeper navigates the boat. The same composition – a woman navigating a boat carrying a beehive and a tree – exists on the seal from Makrys Gialos. Although it is difficult to imagine a tree actually been transported on a boat it becomes more realistic if we consider a dwarf tree *Medicago arborea* and the importance of a swarm loss.

Prehistoric wooden and woven wicker beehives have not been found. But the perishable nature of wood and wicker makes the recovery of such beehives, improbable. In the clay sealing CMS II7,218 from Kato Zakros (fig. 31.22) a number of lidded cylindrical objects with "hands" and "bars" on their surface are depicted which were interpreted by Evans and others as "castellated buildings"[253]. We suggest that they represent woven wicker beehives similar to the traditional ones (fig. 52) with a handle on the top to facilitate

---

[252] Nilsson 1950, 350
[253] Evans 1921-1935, I, 308, fig. 227a. Compare the same buildings on the sealing CMS II7,219 from Zakros reproduced in Evans (as above, fig. 227b). Hogarth (1902, 88) speaks of "five towers of ashlar masonry on a hill"

the lifting of the hives. In front of the objects are posed two eight-shaped (bronze) shields to be used for tanging in case of swarming[254].

Fig. 52. Wicker beehives from Chalkidike, Greece (photo Kyrou 2000, 386)

---

[254] We could also identify with board beehives the objects of Kastelli seal CMSVSuppl 1A, 142 from Chania. In the Kastelli seal, the beehives are placed one above the other in two or three series and above them are depicted *"horns of concentration"* and a male figure that has been characterized as *"god, protector, of the city "* (Hallager 1985). The placement of beehives the one above the other was a usual practice in the antiquity according to the testimonies of Columella (*Rust.* 9.7.3), Varro (*Rust.* 3.16.16) and Palladius (*Opus agriculturae* 2.38), as well in Europe in the 17th c. and until today (Crane 2000a, 405-411 and fig. 32.3e; Vrontis 1939, 196). The number of stacked beehives, as the Roman writers say, should not exceed the three because the apiarist could not then reach the more tally placed beehive. The same number of stacked board hives we see in the representation from Kastelli seal. On top of them stands a figure with a stick recalling the description of Virgil (G. 4.110) *"let the watchman against thieves and birds, guardian Priapus [...] protect them* [the beehives] *with his willow-hook"*. Priapus was regarded as the god protector of beehives (Paus. 9.31.2). For the "horns of concentration" see below.

"Horns of concentration"

Minoans used to place on "altars" or at the base of "sacred" trees, bull horns - one of the best-known symbols of Minoan Crete – called by Evans "horns of concentration"[255]. This form is represented in the ceramic "shrine" model found in the "Loom Weights Basement" in Knossos (MMII)[256] (fig. 53). "Horns of concentration" can be seen in the sealing from Kato Zakros CMSII7,1 of LMIB-II where we see a male figure and further away a female one, depicted in a cavalier perspective, and a swarm of bees hanging from a tree branch (fig. 31.23). In the sealing CMSII6, 3 (fig. 31.24) we observe a woman beekeeper in front of a beehive, a disk shape honey comb inside the beehive, and two pairs of horns placed upon the beehive.

Fig. 53. Miniature "shrine" from Knossos (photo Herakleion Museum).

We think that here too, an ancient beekeeping practice, one that has been preserved up to present days, must be recognized: the placing of a skull of an ox or other horned animal on the beehive in order to exorcise the "evil eye" believed to cause the loss of swarms or the death of the bees. Belief in the evil eye is well nigh universal and it is well known that uncommon and striking objects are frequently used to attract the evil eye and divert it from the object liable to be injured[257]. The evil eye was a widely spread belief in prehistory and in classical antiquity, the Byzantine period and up to modern times in Greece[258]. It is almost certain that the Minoans too, believed in the evil eye given their common belief in magical practices as denoted by the use of talismanic seals[259] and apotropaic amulets[260]. The famous "mythic" charmers Telchines (Callim. fr.75; Ephorus fr.104; Diod. Sic. 5.553) and Idaioi (Hellanicus fr.89; Pherec. fr.7; Diod. Sic. 5.64.4; Strabo

---

[255] Evans 1901, 135; Evans 1921-1935, I, 508, IV, 200-1, 467; Nilsson 1950, 165 ff; Cook 1964, 506ff; Dietrich 1996, 91; D'Agata 1992
[256] Evans 1921-1935, I, 221, fig.166; Nilsson 1950, 87-8
[257] Lykiardopoulos 1981
[258] Prehistory: Thomsen 1992 (Mesopotamia); Dundes 1981, 257ff. Classical Greece: Phylarchus *F. Gr. Hist.* 81 F 79a; Euphor. fr. 185. The oldest written testimony for the evil eye is found in Democritus (ap. Plutarch *Quaest conv.* 5.680-682). The *vaskania* (evil eye) is also mentioned by Plato (*Phaedo* 95.b); Theocritus (*Idyll* 6.34); Dionysius Halicarnassensis (Ἐν Ἀδήλοις 1.6); Heliodorus (Aeth. 4.5); Plinius (HN 7.2); Ovid (*Amor* 1 and *Eleg* 8.15); Virgil (*Ecl* 3.103 the *vaskania* of the animals); Horace (*Epist* 1.14.37); Joannes Chrysostomus (*PG* 61.105-6); *Apocryphal Books* (*Wisdom of Sirach* 14.6-8 and *Testament of Benjamin* 4.2). For a more detailed analysis regarding the belief in the evil eye in the Greco-roman and Islamic world, see Michalopoulou-Veikou 1996, 14-33. For the same belief in the Greek folkloric tradition see Romaios 1955; Lawson 1964, 8-15. Byzantium: Kukules 1948, 244-7.
[259] Onassoglou 1981
[260] As the amulet from necropolis of Hagia Triada (Nilsson 1950, 322, fig. 152)

10.3.22) had close connections with the Kouretes and Crete - the archetypal evil-eye-ers being Telchines (Strabo 14.2.7).

Fig. 54. (1) Representation of Knossos "Palace" with horns on the roofs (design by Piet de Jong) (2) Horns on the roof of a house in Malta used as mean of protection from the evil eye (Zammit-Maempel 1968)

The harming effect of the evil eye on humans, houses, crops and animal products is a belief still prevalent in most parts of the world. In Anakou, Cappadocia, a donkey cranium hanging on a tree branch was used as an evil eye protection for crops[261]. The same practice is reported in *Scholia in Aristophanem Plutum* (ΣAristoph. 943). A pig head was used in a similar way in modern Crete[262] while a cow head hanging from a tree is depicted on a wall painting from Pompeii[263]. But the most common means of protection against evil eye was a horned animal skull[264]. Besides humans, animals and crops, evil eye could affect whole cities as well as private houses. The ancient Egyptians used to cut the head of an ox and throw it in the river in order for all the curses of the city (caused by the evil eye) to fall on it and leave the city (Hdt 2.39). A similar

---

[261] Kostakis 1963, 286
[262] Psilakis 2005, 433
[263] Evans 1901, 128
[264] Elworthy 2004, 181-233; Rouse 1896 (Lesbos, Greece); Blackman 1910 (Egypt); Hornell 1925 (Madeira); Zammit-Maempel 1968 (Malta, Corfu, Arabic countries); Megas 1941, 83 (Epirus, Greece)

ritual, the Bouphonia, existed in ancient Athens[265]. According to the tradition it was introduced by Erechtheus who also introduced apiculture (Columella *Rust*. 9.2). Horned heads of animals on house terraces and facades were used as protective means against the evil eye until recently in Arabic countries, in Malta (fig. 54.2) and in Greece[266]. We believe that the large stone horns in the courtyard at Nirou Chani, the balustrade of the north porch at Gournia and near the north and south entrances to the palace at Knossos[267] (fig. 54.1), had a similar functionality – to protect from the evil eye. In some rings, frescoes and pyxis representations also, Minoan buildings are crowned by horns[268] recalling the apotropaic symbols sculptured on building façades in Assyria[269] and in modern Greece. Traditional apiarists too, paid particular attention to the evil eye and had devised many means to avoid its consequences. In Greece (Roumele) and elsewhere (Cappadocia, Pontos), the honey harvest was done only during the night to avoid the evil eye[270]. Perhaps for the same reason the gathering of honey in antiquity was done at night (Plin. *HN* 11.15). Beekeepers in Aitoloakarnania, in Greece, used to place on the hives a skull of a donkey[271] and in Rhodes they used to hang skulls of big animals (pig or donkey) on the trees near the hives[272]. The six animal heads depicted in the Mycenae CMSI,17 ring (fig. 31.4) could represent such a prehistoric Aegean belief. In Cappadocia (Anakou) they used a tortoiseshell placed on the hives[273]. But in many places the skull belonged to a horned animal, as in Greece, in Cyprus[274] and in the Caucasus[275] (fig. 55). Accordingly, we believe that the horns on Minoan beehives were just another apiculture related item, a charm, one of the numerous that have been preserved up to modern times by peasants and beekeepers.

The same apotropaic qualities we should probably attribute to tanging, a very ancient bee charm[276]. Although we know today that bees can hear certain frequencies, a satisfactory scientific explanation of whether tanging is effective against the departure of a swarm, is still lacking[277]. Aristotle (*Hist. An.* 627a.17) doubted that bees are capable of hearing while wandering for the popularity of tanging in his time. A probable explanation is that the impressive sound of bronze concave objects could distract the regard of the evil-eye-er from the swarm. We know that in antiquity as in modern Greece the sound of bronze

---

[265] According to a hypothesis by Harrison 1903, 111, n2. It is believed that Bouphonia (also called Dipoleia) was a festival of pre-Bronze Age origin (Simon 1983, 105).
[266] Zammit-Maempel 1968. Apotropaic symbols sculptured on buildings façades were widely used until recently in Greece (Konstantinopoulos 1987, 61)
[267] Rehak et al. 2001, 435; Evans 1921-1935, II.2, 589 fig. 367; Soles 1991, 46-7
[268] As for example the "Pillar Shrine" fresco, the stone rhyton fragment from Gypsades, ivory pyxis from Agia Triada, seals from Knossos etc
[269] *Bit Meseri* series from Assyria (Meier 1941/44)
[270] Karalis personal communication 2004 ; Topalidis 1968/1969 ; Kostakis 1963, 388. In Crete (Ionochorion) they believed that the bees themselves drowned the person with the evil eye during the honey vintage season (Megas 1941, 114)
[271] Loukopoulos 1983, 396
[272] Vrontis 1939, 207. A cow head hanging from a tree can be seen on a coin of the 5th c. BCE from Caulonia in Italy, on which often exists the symbol of the bee as well (Noe 1958, 128).
[273] Kostakis 1963, 286
[274] Rizopoulou-Igoumenidou 2000, 399 and n 5.
[275] Armbruster 1926, 23 reported by Ransome 1937, 111, n 1
[276] See Part 1 "Beekeepers and beekeeping practices"
[277] Crane 2000a, 567

bells was used as an amulet[278]. Little bronze bells were worn on babies against the evil eye in the time of St. Chrysostomus (*In epistulam i ad Corinthios* 61.105.55) who reports this custom only to condemn it as an ancient pagan remnant. Small bronze bells around animals' necks were a protective mean from evil eye influence on the herd[279]. Mycenaean small eight shaped shields worn as gems were probably used as amulets too[280].

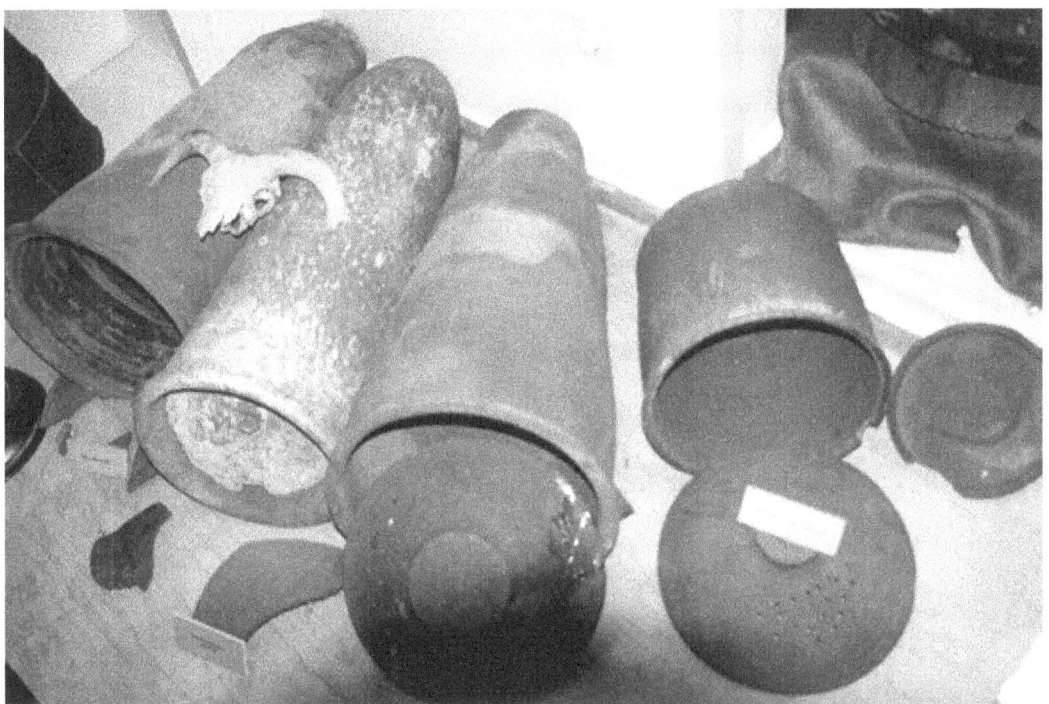

Fig. 55. Traditional beehives from Greece with a goat head on them for evil eye protection (private collection of T. Bikos, photo P. Kamilakis)

---

[278] Polites 1975, 206-7 with references
[279] Aikaterinidis 1996 s.v κουδούνι. Ceramic "sheep bells" have been found in Minoan Crete. Two interpretations have been offered: one identifying them as sheep bells, largely because of the perforations at or near the top of the figurines that are thought to be for suspending a clapper, and the other identifying them as horned masks, largely because of a painted faience bell from Poros (Evans 1921-1935, I, 175; Nilsson 1950, 191-3; Platon 1948, 833-46)
[280] Danielidou 1998, 101

"Sacral knot"

Another well known iconographic form - a knot - appears in the sealing from Hagia Triada CMSII6,4 (fig.31.11). Evans hails the knots as sacral[281] but Nilsson questions this belief[282]. A knot in ancient and modern belief is universally considered as a magical mean that "ties up" the free deployment of a natural process and its presence impedes whatever such natural process goes on nearby[283]. It is possible then, that in prehistoric Aegean beekeeping the knot was used as a means to prevent the swarm from flying away, thus explaining its appearance in the beekeeping scene of Hagia Triada CMSII6,4 sealing.

"Temenos"

A fragment of a steatite pyxis from Knossos shows an "altar" built with square blocks (fig. 56). A pair of "horns of consecration" stands on it. Around the "altar" there is an enclosure wall which is considered to be the wall of a "temenos", a wall that delimits a "sacred" place[284]. It is generally accepted that in the right side of Knossos bronze ring CMSII3,15 (fig. 31.17) (in which we have previously identified dot-bees entering a ceramic beehive and in which we can also recognize a woman beekeeper surrounded by bees and some swarms on branches) is also depicted a "wall enclosure" on which we can distinguish three pairs of horns. We find that the similarity of this "wall" with the "altar" on the Knossos pyxis could not be overlooked. We believe that both are brick beehives with pair of horns superimposed[285]. The enclosure wall on the pyxis could represent the practice of the Minoan beekeepers to put their hives inside tall enclosure

Fig. 56. Pyxis from Knossos

---

[281] Evans 1925-1931, IV, 608

[282] Nilsson 1950, 163-4

[283] For an analysis of the universally accepted magical properties of knots see Frazer 1936, 293-317. For Greek lore see Aikaterinidis 1996 s.v κόμπος. For the magical use of knots in ancient Egypt see Moret 1908, 249-50, Erman 1909, 181-2, fig.99. For ancient Mesopotamia see Contenau 1937, 172. In ancient Greece the "knot of Herakles" the *nodus Herculeus*, was endowed with magical virtues. The knot, like the horns, is also used as an amulet against the evil eye. For reports from modern Symi see Polites 1975, 256 and from Crete see Chrisoulaki 1958, 392

[284] Evans 1901, 101-2, fig. 2; Nilsson 1950, 120. For "sacred enclosures" in the Prehistoric Aegean see Rutkowski 1986, 99-118

[285] The existence of brick beehives in antiquity is attested by Columella (supra)

walls similarly to the Romans (*alvearium, mellarium, apiarium*) (Columella *Rust.* 9.5.2) and modern traditional beekeepers[286].

Fig. 57. *(1) Apiary wall enclosure in modern Crete (Rackman and Moody 2004 fig. 12.9) and (2) in Roya valley, France (photo E. Crane from Crane 2000a, 325, fig. 32.5c)*

Bees' enclosure provides shelter from the wind, which bees dislike (Verg. *G.* 4.8). In modern Sphakia, in Crete, people built beehive enclosures at the top of a set of agricultural terraces, or on uncultivated slopes where thyme bushes are abundant (fig. 57.1-2). Frequently there is a tree growing inside or near the enclosure[287] (to prevent the swarm to fly away) just like the tree behind the wall represented on the steatite pyxis from Knossos.

---

[286] Loukopolos 1983, 391-2; Kukules 1948, 348. For reference to European apiaries enclosed by a wall see Crane 2000a, 324-5, fig. 32.5a, 32.5b, 32.5c. For Crete see Nixon 2000; Rackman and Moody 2004, 218, fig. 12.9
[287] Oikonomidis 1965-1966, 632

"Double axe"

Among the scenes which we analyzed in an apiculture context the *labrys* (double axe) appears only in the ring from Mycenae CMSI,17 (fig. 31.4). We have so far explained the other motifs in such a context so we will attempt to provide an analogous interpretation for labrys as well.

The labrys of the CMSI,17 ring according to Mylonas is shown, in perspective, standing on the ground[288]. Double axes stands have been found in Hagia Triada, Mycenae and Knossos. Why labrys was set up on poles? Why in a sealing from Hagia Triada (HM 592), a woman and another person are shown to carry double axes as if they were labarum (gr. *λαυρᾶτον*)? Was labrys an emblem and what did actually represented? Emblems, frequently, are borrowed from the realm of animal world: the owl, the crow, the griffin, the fox, the serpent, the dolphin, the cock (the emblem of Idomeneus, the decedent of Minos) are some examples. We believe that labrys is the schematic representation of a bee. In the handhold, we see the body of a bee while its symmetric blades portray the wings of a bee (fig. 58).

Fig. 58. Hypothetical origin and evolution of the schematic meaning of the double axe (drawing V. Harissis)

The frequent variant with a double blade at each side as in CMSII7,7 or in the small ivory double axes from a sanctuary from Zakros[289] (fig. 59.1), the golden jewels from the fourth shaft grave of Mycenae[290] and the basket-shaped vessel from Pseira (fig. 59.6), corresponds to the two pairs of wings that the bee actually has (fig. 58). As for the diagonal and distressed lines that we often see on the blades of the double axe, and for which Evans claimed a special religious importance[291], while other supported that they symbolize "lightning"[292], we believe that in the naturalistic Minoan art they could very well represent the "veins" that exist on the wings of the bee (fig. 59.2-3). The double axes on a pithos fragment form Psychro (1400 BCE) and on the basket-shaped vessel from Pseira depict

---

[288] Mylonas 1966, 149-50, 157, 169-71
[289] Platon 1971, 131
[290] Cook 1965, II, 538, fig. 409 c,d
[291] Evans in *Ann. Brit. Sch. Ath.* 1900-1901 vii, 53 fig 15 reported by Cook 1965, 639
[292] Cook 1965, 640

schematically the head of the insect (fig. 58; fig. 59.4; fig. 59.6) just like the double axes on the Hagia Triada jug (fig. 59.5) which also depict the body of the bee.

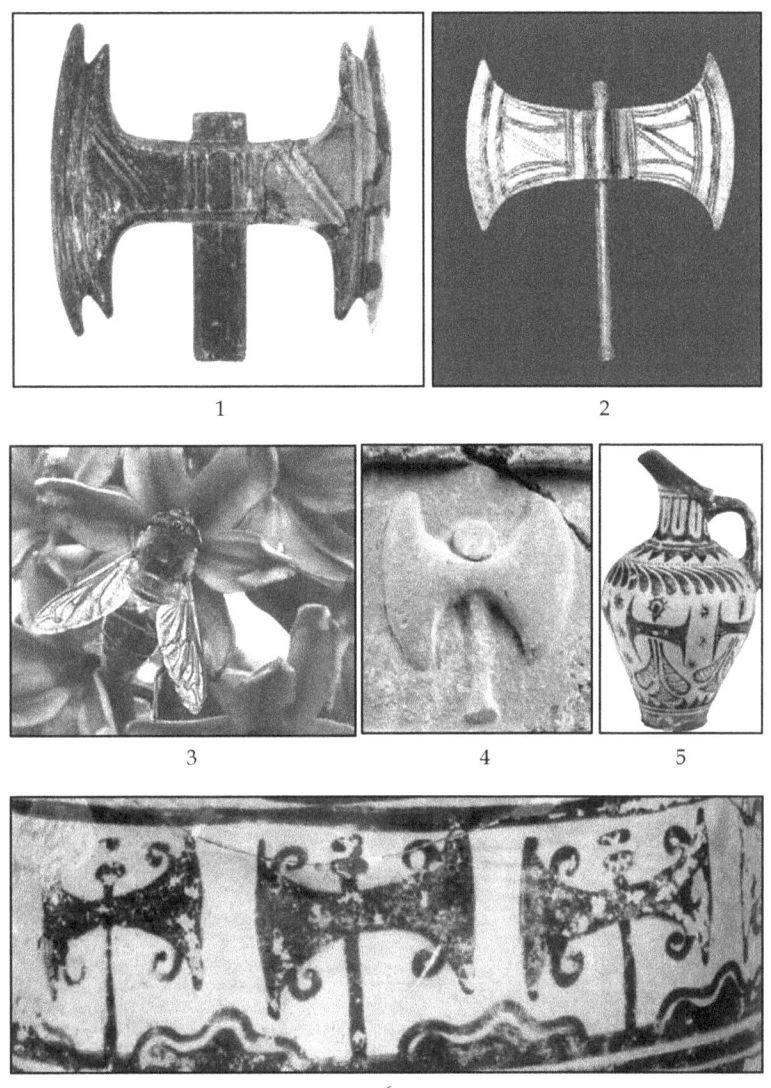

***Fig. 59**. (1) double axe from Zakros (2) double axe from Arkalochori (3) wing veins of a bee (4) double axe on a pithos fragment from Psychro (5) double axes on Hagia Triada jug (6) double axes on a basket-shaped vessel from Pseira (photos 1,2,4,5,6 Herakleion Museum)*

When double axes bear decorations on them, the motifs come from the realm of plants and insects (as the lilies and "butterflies" on the double axes from Zakros and the flowers of the double axe on a vase from Mochlos[293]).

The bee was used as a title by the Pharaos of Lower Egypt; a bee was the emblem of the city of Delphi and Ephesus and was used as an insignia by the Frankish king, Childeric I (died 481) and Pope Urban III in 1626[294]; a honey bee was a prominent political

---

[293] Nilsson 1950, 208, fig. 102
[294] The bee was used as the insignia of the Italian family of Barberini in which belonged the Pope Urban III.

symbol in the empire of Napoleon Bonaparte; several old French, English and Italian families used the bee as an insignia; the bee appears as a symbol of organizations, as the Estates of Burgundy in 1845 and the "Order of the Bee" founded by Louise Benedicte, wife of Louis Auguste de Bourbon, Duc du Maine in 1703, and others[295]; the Mormons established their state of Desert in Utah by embracing the beehive symbol to represent their goal of an industrious, communal, society; honey bee is designated as state insect in seventeen North American States[296]. Many other examples can be given but our point is that the symbol of the bee is probably the most widely used insect in heraldic charges and this for a simple reason: bee symbolizes the spirit of diligence, stable government, good management and organization, all the qualities found in a beehive, that man wishes to apply on his society. Maybe, for similar reasons, labrys, being a representation of bee - an object of human fascination and delight for eons – was an emblem in prehistoric Aegean societies as well.

---

[295] Ransome 1937, 234
[296] Horn 2005

Epilogue

In the second part of the monograph we thoroughly analysed a total of 17 out of the 19 motifs/forms appearing in 12 Mycenaean and Minoan rings (11 gold ones and a bronze one), 2 seals and 10 sealings[297]. By applying a decipherment grid based on an apiculture context we have been able to obtain a coherent interrelation - a coherent "grammar" of motifs/forms - and to identify practices and paraphernalia of an apiculture whose existence and wide propagation in prehistoric Aegean we have shown in the first part of this monograph. In the first part, this "apiculture" approach has permitted us to accurately identify a lot of the so called "sacral objects" from Knossos with simple beekeeping paraphernalia. Similarly, in the second part, we have been able to show that the "fruits on sacred trees" are swarms of bees on trees, the "festoons" and "pearls" are flying bees, the "ritual bending of the sacred tree" is a scene of swarm capture by the people engaged in apiculture activities usually called "flying gods" and "great goddesses", while the so called "beatyls" and "altars" are simply different types of beehives (stone, wood and wicker beehives) placed in beekeeping enclosures ("temenos"). The "bird epiphanies of deities" are actually depictions of birds preying on bees, while the "sacral knot" and the "horns of concentration" are primitive means of protecting the beehives from the evil eye still in use by modern beekeepers. Finally, the "double axe" is simply the schematic representation of a bee used as emblem by the beekeepers on their seals and vessels.

The 24 rings, seals and sealings analyzed here are but a small fraction of the approximately 4500 rings, seals and sealings of the Aegean Bronze Age corpus and represent only about 3% of the total corpus that depicts human figures, but they also constitute some of the most well known and most frequently discussed rings, seals and sealings that inspired, since Evans, generations of scholars to contextualize and to explain on the basis of their imagery an Aegean prehistoric religion. Our interpretation questions such a theory and we prefer to see on the rings practices that we identified with apicultural scenes.

We believe that the signet rings and seals in question were used by officials-overseers responsible for the control of the highly valued beekeeping products in a supraregional network that linked the main Minoan and Mycenaean sites in an exchange of prestige goods[298]. The fact that the majority of the presented rings were made of gold, reflects exactly the lucrative nature of honey and beeswax commerce; both were sought-after commodities, and commanded high prices.

---

[297] We deliberately omitted an analysis of the so called "floating ray" (appearing in the Mochlos ring) and the "floating eye" (appearing in the AM 1919 ring) because we believe that their considerable lack of clarity due to their small size renders them highly ambiguous (but cf Kyriakidis 2005). However, we believe that they could represent flying swarms

[298] Schoep 1999, 217; Laffineur 1990; Laffineur 1992; Sakellariou 1999. Similarly the MMIB-MMII seals showing looms weights indicate the administration of textile production (Burke 1997) and the seals with domesticated animals could indicate the livestock transactions known from Linear B tablets

Whether these beekeeping scenes had also a religious dimension of a "bee-cult" remains to be shown[299]. Established ideas about the prehistoric Aegean religion should be reconsidered, thus creating significant challenges for future research.

---

[299] Several scholars have proposed the existence of a bee-cult in prehistoric and classical Greece: Cook 1895, Harrison 1903, 442-3, Neustadt 1906, Marconi 1940, Picard 1940, Willetts 1977, Somville 1978. However, their arguments are based exclusively on a few scarce literary sources of Classical, Hellenistic and later periods (Aesch fr. 87, Didimus ap. Lactantus *Div. Ins.* 1.22, Pind. *Pyth.* 4.60, Mnaseas fr. 3.50 Apollodorus Περί θεών FGrH Jacoby II B No 89, p 1045; Aristotle *De generatione animalium* 761a, 5

Appendix

Virgil's Aristaios: an ancient beekeeping educational myth

Aristaios (Ἀρισταῖος, Aristaeus), was a rustic deity credited with the discovery of many rural arts, including beekeeping. An early source (Bacchyl. fr. 62 B. III 587 = fr. 45 Blass–Snell) records four possible parentages of his: i) Karystos (son of Cheiron and the nymph Chariklo, daughter of Apollo), ii) Cheiron, iii) Gaia and Ouranos and iv) Apollo and Kyrene. Most ancient sources consider Aristaios as a son of Apollo and Kyrene, daughter of king of Lapithai, Hypseus (Hes. fr. 215, 217; Pind. *Pyth*. 9.5-70; Phylarchus fr. 16; Paus. 19.17.3; Diod. Sic. 4.81.1; Hyg. *Fab* 161, Hyg. *Poet. astr.* 2.4, Ov. *Fast.* 1.363, Verg. *G.* 4.320, Nonnus *Dion.* 5.212). Apollo saw Kyrene in the vales of Pelion, wrestling with a lion and fall in love with her. Spurred on by the approval of Cheiron, Apollo persuaded the maiden to follow him from Thessaly to Crete and then to Libya where the children Aristaios and Autouchos were born (Agroetas *Λυβικά* fr. 2 M. IV 294; Pind. *Pyth*. 9.5-70; Acesander fr. 4; Phylarchus fr. 16). Apollo entrusted Aristaios to Hermes (Hes. Fr. 217) who took him back in Thessaly to the Horai and Gaia (Pind. *Pyth*. 9.85) or to Cheiron and to the Muses *"on the Athamantian plain in Phthia, round Mount Othrys and in the valley of the sacred river Apidanos"* (Ap. Rhod. *Argon*. 2.506; Diod. Sic. 4.81.2;).

Aristaios married in Boeotia, the daughter of Kadmos, Autonoe, by whom he begot Aktaion (Apollod. *Bibl*. 3.30; Callim. *Hymn* 5.108; Paus. 10.17.3; Diod. Sic. 4.81.1, Hyg. *Fab* 181, *Poet. astr*. 2.4, Nonnus *Dion* 5.212) who was torn to pieces by his own dogs on the mountain of Kythairon (Apollod. *Bibl*. 3.30). After the death of his son and having assembled in Arcadia the Parrhasian tribe, the descendants of Lykaon, Aristaios was summoned to Keos (from Phtia or Boeotia according to sources) on his father's Delphic oracle command. At that time the island – like other *"islands of Minos"* – was scorched during the dog-days of summer (Ap. Rhod. *Argon*. 2.506). There he built an altar to Zeus Ikmaios (Zeus of "Moisture") and sacrificed a bull (Nonnus *Dion*. 5.273) to the Dog-Star (Κύων, Μαῖρα, Σείριος, Sirius) so that Zeus would send the cooling Etesian winds (Ap. Rhod. *Argon*. 2.506; Call. *Aitia* fr. 75.32-37; Hyg. *Poet. astr*. 2. 4; Diod.Sic. 4.82.2). He then returned to Libya and from there he sailed to Sardinia (Paus. 10.17.3) where he begot two sons, Charmos and Kallikarpos (Diod. Sic. 4.81.1). Later he moved to Sicily (Diod. Sic. 4.82.5) and in the end of his life he retired to Thrace (Diod. Sic. 4.82.6). Pindar (fr. 251 Snell) and Servius (ap.Verg. *G*. 1,14) record Aristaios' reign in Arcadia - Virgil calls him *"Arcadian master"* - and Nonnus his participation in the war of Dionysus in India. According to Oppian (*Cynegetica* 4.265) Aristaios in a certain point of his life, lived in a cave in Euboea where he begun the pedagogue of Dionysus. Pliny (*HN* 7.199) calls him an Athenian.

Although Aristaios was also concerned with music, healing and prophecy (Nonnus *Dion*. 17. 357 ff) he was primary the herdsman-hero par excellence. Aristotle (*Mir*. 838b) refers to him as γεωργικότατος, and according to Herakleides (*FHG* II 214.9.2) he was a shepherd who learnt the care of sheep and cattle from the Nymphs. Apollonius (*Bibl*. 4, 1132-3) and Cicero (*Nat. D*. 3. 18) credit him with the discovery of olive-oil, and Diodorus

(4. 81-2) states the he was taught by the Nymphs how to make cheese and cultivate the olive. Pliny (*HN* 7.56.199) attributes to Aristaios the invention of the olive-press. In Sicily he was honored as a god by olive-growers (Diod. Sic. 4.82.5). According to Suda (s.v σίλφιον) Aristaios was credited with the cultivation of *silphium*, a North African extinct plant species of the genus *Ferula*, whose gum or juice was prized by the ancients as a medicine and a condiment[300]. Nonnus accounts for Aristaios' inventions and discoveries of hunting, snaring, tracking, the hunter's high boots and short tunic, olive-oil and herdsmanship.

But Aristaios' main contribution was beekeeping, an important branch of agriculture in antiquity. Apollonius (*Bibl.* 4. 1128) calls him *"the honey-loving shepherd who discovered the secret of the bees"*. He was taught the art of apiculture by female deities, the nymphs "Vrisai" or "Vlisai" (Arist. fr. 511; Diod. Sic. 4.81.2; Oppian *Cynegetica* 4.271), a name that we should probably etymologise from the verb βλίττειν, "honey harvesting" (Pl. *Rep.* 8.564e) or the Cretan word βριτύ for "sweet" (Hsch. s.v.). Nonnus (*Dion.* 5.247-9) says that it was he who first wore protective clothes for the bees, he smoked with a torch his bees to pacify them, he used tanging to attract the swarms, honey for medicinal purposes, and used the first beehives made from trunks of oaks which he carried from Cyrene (Nonnus *Dion.* 5.212). Such beehives were in use until recently in Libya which in antiquity (Hdt 4.193; Plin. *HN* 11.14) and up to today was famous for its apiculture[301]. Oppian (*Cynegetica* 4. 265 ff) tell us that "*Aristaios . . . instructed the life of country-dwelling men in countless things . . . he first brought the gentle bees from the oak and shut them up in hives . . .* [he lived with] *the Nymphai that have bees in their keeping.*" According to Pliny (*HN* 14.6; 7. 57) Aristaios invented honey and mixing of wine with honey.

But Aristaios memorable appearance as a beekeeper is in Virgil's fourth *Georgic*. Here we are informed that his father Apollo was surnamed "Thymbraios". This name refers, in our opinion, to the plant *thymbra* (*Satureja thymbra*, savory), a plant that belongs to the family of *Lamiaceae* (also known as mint family) along with rosemary, origanum and thyme considered as some of the most important bee plants (Verg. *G.* 4.8). Thymbraios Apollo, in modern Greek, would be called "riganas" (man of origanum) the exact name used in Central Greece (Aitolia) for the saint protector of beekeepers and apiculture[302], a role that Apollo was credited in the *Homeric Hymn to Hermes* (552-566). In Latin the name of origanum is *satureia*, that is to say the plant of Satires who according to Ovid (*Fast.* 3.735) discovered apiculture. Thymbra was also the name of a city of Troy that was founded by Dardanos, whose name refers to δάρδα (darda), meaning the bee (Hsch s.v. before the "correction" by Latte).

Virgil's myth also contains two interesting accounts which we will now discuss in the following.

---

[300] Silphium once formed the crux of trade from the ancient city of Kyrene. It was so critical to the Cyrenian economy that most of their coins bore a picture of the plant (Tatman, 2000)
[301] Newberry 1938
[302] Loukopoulos 1983, 393

When the bees of Aristaios died suddenly, he went to consult his mother in Tempe, to the river Peneus (*G.* 4.317). She referred him to the seer Proteus, for an explanation for the death of his bees. Proteus spent part of his time on the islet of Pharos near the Delta of the Nile in Egypt (or in the Carpathian Sea near Crete) and part at Pallene in Chalkidiki[303]. From Proteus, Aristaios was told that the illness of his bees came as a punishment for his attempt to rape Eurydice, the wife of Orphee.

We believe that this represents an ancient testimony of a very old rule imposed on beekeepers, that of sexual abstention for one or more days before the opening of beehives in spring or before the vintage of honey in summer (Columella *Rust.* 9.14.3; Plut. *Conjugalia praecepta* 144D; Palladius *Opus agriculturae* 1.37.4, 4.15.4; *Geoponica* 15.2.19)[304]. This rule of sexual abstention exists today among the traditional apiarists in Greece but also in primitive tribes[305] and undoubtedly, it is connected with the natural repulsion that the bees feel for certain odours, a fact that many ancient writers testify (Arist *Hist. an.* 626a; Ael. *NA* 5.11; Verg. *G.* 4.48; Varro *Rust.* 3.16.6; Plin. *HN* 11.15). Before the honey harvest, the beekeeper had to abstain from strong-smelling foods and salves, must not get drunk, and had to cleanse himself thoroughly. The apiarists of Egypt, in 3rd c. BCE, shaved their forehead so that they do not smell of grease, an odour that the bees dislike most (Aristophanes Gramm. *Historiae animalium epitome* 1.37). Even modern apiarists recommend particular assiduity in odours when they open the beehives[306]. It has been scientifically proven that worker bees are 10 to 100 times more sensitive than man to wax, flower and other odors biologically significant to them. This highly developed olfactory system is used by the bees for various functions of vital importance as foraging and communication[307]. It is thus possible that alien odors can perturb their functions and lead to hostility, disorientation and swarming. Aristaios by not respecting the rule of sexual abstention, involuntary provoked the escaping of his bees.

We should relate with the above rule, the myth of Rhoikos, a Cnidian who having seeing that an oak was about to fall, he had it propped up (Charon of Lampsakos fr. 262 F 12a = Σ*Apoll. Rhod.* 2.476-483; *Etym. Magn.* s.v. *Αμαδρυάδες*, Σ*Theocr.* 3.13). The "nymph", who had been doomed to perish with the tree, acknowledged her debt to Rhoikos and bade him ask for whatever he wished. When he asked to have sexual intercourse with the nymph (συγγενέσθαι), she told him to avoid relations with other women and said that a

---

[303] Callimachus (fr. 254) calls Proteus a seer from Pallene (modern Kassandra in Chalkidiki), an attribute also attested by Lycophron (*Alex.* 126-7) and Nonnus (*Dion.* 43.334). Homer (*Od.* 4.349-570), Herodotus (2.112-20) and Lycophron consider him Egyptian while Virgil calls him native of Pallene who sojourned also in Egypt

[304] The same practice applies for wine vintaging in Kythnos (Megas 1941, 115). Burkert (1987, 60, n.12) cites the sexual abstinence rule applied before hunting in the epic of Gilgamesh

[305] In Rhodes (Vrontis 1939, 205) and in Naxos (Oikonomidis 1965-1966) the apiarist should abstain from sexual intercourse the day prior to the vintage of honey. Many "primitive" tribes today have the habit of sexual abstention the day before the honey hunting (Ransome 1937, 287-288; Leibovici 1968, 37-8)

[306] Bikos 1999b, 106. Traditional apiarists of the region of Preveza in Epirus, Greece, avoid eating garlic and fish before opening the beehives (we owe this information to Mrs Karampini, an apiarist)

[307] Winston 1987, 16-8 with references

bee would act as a messenger between them. The bee flew by while Rhoikos was "playing draughts" (πεσσεύοντος), but he for some reason spoke crudely and angered the nymph, who "maimed" him (πηρωθῆναι - also with the meaning of mutilate, castrate). If we translate the word πεσσεύοντος with the metaphorically meaning of gamble we can udenrstand that Rhoikos gambled/jeopardized, by attempting adultery, his great fortune to have gained the nymph as his wife. It is then possible to reconstruct the myth as follows: Rhoikos asked the nymph to marry him. The nymph accepted and assigned to a bee to watch him. When he tried to commit or actually committed adultery, he was punished. This specific intermediation of the bee implies its capability to sense the act of sexual contact, or the hormones that are secreted before and during the act. According to Pindar (fr. 252 SCHROEDER = Plut. *Quaestiones Naturales* 36) Rhoikos was stung by the bee because of his infidelity since *"bees are quicker to sting those who have recently committed adultery"*.

The second motif in Virgil's account refers to the act of *bougonia*. After the bees of Aristaios were suddenly lost, his mother Kyrene advised him to establish altars, sacrifice cattle (four male and four female oxen) and leave their carcasses exposed. From the carcasses, new swarms of bees rose (Ov. *Fast*. 1.363; Verg. G. 4.316-566). Thus she revealed to him the secret of *bougonia*. This in Virgil's account takes place in Thessaly but many indirect clues show that Aristaios executed the bougonia in Kea. Our belief is that Aristaios imported apiculture in Kea by the means of bougonia - Nonnus (*Dion*. 5.273) tell us that Aristaios scarified a bull in Kea - so that he would end the famine that plagued its residents. The scholia in Apollonius (ΣApoll. 169.12) mention that Aristaios was the first to introduce apiculture in Kea while Eumeros (ap. Columella *Rust*. 9.2.4) says that apiculture was first discovered in Kea.

In antiquity it was a widespread belief that bees were born from a dead ox or a dead bull (Nic. *Ther*. 741; *Alex*. 447; Arist. *Gen. An*. 759a). The process was called bougonia (βουγονία) and the oxen-born bee βουγενής, βουγόνας, ταυροπάτωρ, βούπαις or βουποίητος (Porph. *De antr. nyph*. 1 5.8, 18.9; Varro *Rust*. 2.5; Theocr. *Syrinx* 3 and ΣTheocr.; Erycius *Anth. Pal*. 7.36; Straton *Anth. Pal*.12.249; Meleagros *Anth. Pal*. 9.363; Vianor *Anth. Pal*. 9.548; Suda s.v. βούπαις). The term "oxen-born" is first mentioned in the 4th c. BCE by Callimachus from Cyrene (fr. 383,4 ap. Hsch. s.v. βουγενέων). Archelaus (fr. 128) from Egypt (3rd c. BCE) also reports the same belief and he calls the bees *"children of the dead ox"* (βοός φθιμένης πεποιημένα τέκνα). However, this belief must be even earlier since Eumelus the Corinthian in 8th c. BCE had written a work that was named Βουγονία (Varro *Rust*. 2.5.5; Euseb. (Hieron) *Chron*. 01.5.1). Many other writers report the effectiveness of the bougonia process without providing any further details (Ov. *Fast*. 1.377, *Met*. 15.364; Plin. *HN* 11.22.70, 21.47.81; Ael. *NA* 2.57; Serv. *Georg*. 4.286; Lib. *Extr* 1.1.90; Isid. *Orig* 12.8.2; Sext. Emp. *Pyr*. 1.42; Philo *De Speliabus legibus* 1.291.4; Porph. *De antr. nymph*. 18.9; Simpl. *In Phys*. 9.239.18; Joannes Philoponus *In Aristotelis physicorum libros commentaria* 16.107.14; Origen *C. Cels*. 4.57). The bougonia belief is also reported by Persian sources[308] as well as in the *Yerushalmi Talmud* (*Shabbath* 1.3b) and the *Babli Talmud* (*Baba Qamma* 16a) of the 5th and 6th c. respectively. In 1474 Petrus de Cresentiis in *Ruralia Commoda* also refers to the method. In the 14th c., Konrad

---

[308] Evans A. 1884. *Archaeologia* 48:23, reported by Ransome 1937, 277

von Megenberg, writer of the first German book of natural history (*Das Buch der Natur*), claimed that the bees are born from the skin and the stomach of the ox. Michael Herren (1563) gives a detailed description of bougonia drawn from *Geoponica*. Johannes Colerus (1611) whose book *Oeconomia ruralis and domestica* constituted the book of reference for many generations of apiarists, expresses the same belief in bougonia[309]. The method appears even in European apiculture books of the 18th c.[310].

A most detailed description of the bougonia process can be found in *Geoponica* (15.2.21)[311]. There, it is reported that the best bougonia method, contrary to the instructions of the king Juba of Libya, (1st c. BCE), who recommended the use of a wooden "larnax", is to build a brick room with dimensions 4.5 by 4.5 metres with a window in each wall, and a door. In the room, a two and a half years old calf should be placed, and certain young men should beat it to death while exercising care to keep the skin intact. Following this, the natural apertures of the animal are sealed with thin buckram infused with bitumen. Thyme and laurel (two of the most important bee plants) are dispersed in the area, and finally all openings of the room are sealed hermetically with clay. On the second week, air was let into the room and then it was sealed again. After ten days, the room has filled with swarms while the horns, the bones, and hairs, are all that remain of the calf. The queen bee (the king in the original text) emanates from the skull of the calf while the workers emerge from the flesh (15.2.30). In the hermetic *Cyranides* (2.39.32) it is reported that from the dead bull worms are born after one week, and bees after three weeks, the same time frame that is reported in *Geoponica*. Only Virgil (*G.* 4.308-314) speaks for a duration of nine days. Virgil (*G.* 4. 306) says that the bull sacrifice should be performed when Zephyr (western wind in the beginning of spring) starts to blow and "*when the swallows build their nests and when the fields change colour*", i.e. in the beginning of spring (21 March). Columella (*Rust.* 9.14.6) however, says that according to Democritus, Mago and (paradoxically) Virgil, bougonia was performed between the summer solstice (25 June) and the morning rising of the Dog Star (28 July). It is peculiar that Columella refers to Virgil since Virgil reports that the bougonia ritual was performed in the beginning of spring. The most likely explanation, is that Virgil has confused the name of the "Annual" winds (*etesian* winds), which blow during the period of the rising of the Dog Star, with Zephyr that blows in the beginning of spring. Indeed Virgil himself (*G.* 425-8) places Aristaios' visit to Proteus at the period of the rising of the Dog Star, and consequently the act of bougonia of Aristaios, which followed immediately, cannot be at springtime[312].

The question then arises of how to explain this over a millennia lasting belief in bougonia? The most cited reasoning is the one stating that in antiquity they mistook the drone-flies (*Eristalis tenax*), who's larvae feed on decaying organic material, for bees[313]. But this explanation is rather improbable since the ancient as well as the 18th c. apiarists

---

[309]Ransome 1937, 153-154
[310]Hartlib 1655 and *Dictionarium rusticum et urbanicum* 1704 (both reported by Crane 2000a, 581). Other examples are reported by Ransome 1937, 205
[311]For critical presentation of the sources of bougonia belief see Gow 1944
[312] For a different explanation for this disaccord in Virgil (attributed to poetic reasons) see Mynors 1994, 299
[313] Osten-Sacken 1894

would promptly understand that these flies do not produce honey. The best explanation in our opinion, agreeing with other scholars, is that the belief emanates from the observation that the wild bee *Apis mellifera* finds resort to any cavity, and in particular cavities of trees and rocks, but also in skulls and in thoracic cavities of large animal carcasses[314]. Shakespeare knew this well as he says in *Henry IV* (B, IV, 4.79) "*the bee doth leave her comb, in the dead carrion*". Thus we can explain the bees which Samson saw coming out from the dead body of a lion (*Judges* 14.8.1)[315], the swarm inhabiting the skull of a dead person witnessed by Herodotus (5.114) and the bees emerging from the dead body of a priestess of Demeter (Serv. Verg. *Aen.* 1.430). The observation that wild bees find shelter in dead bodies of large animals must be very old and emanate from dry regions where the species *Apis mellifera* exists, lacking more appropriate nesting sites such as trees and caves like the region of the Delta in Egypt. Indeed, most of the writers admit that the Greeks had acquired this belief from Egypt or Libya (the term Libya sometimes refered to the whole northern Africa as in Herodotus 4.42). Virgil determines the origin of bougonia belief in Egypt, information that was probably drawn by the Latin translation of Carthaginian Mago's lost agricultural treatise that included a chapter entitled "*Getting bees from the carcass of a bullock or ox*" (Columella *Rust.* 9.14.6.)[316], and King Juba of Libya supplied another recipe from the same continent. Democritus may have seen the practice during his journeys in Egypt, as might be the case with Philetas who was tutor to Ptolemy Philadelphus, while Callimachus was a native of Libya (Cyrene) as Aristaios himself. Perhaps the fact that the Egyptian buried the oxen leaving the horns off the ground (Hdt. 2.41.14, Antig. Car. 19), is related to the bougonia belief and more specifically with the belief that the "king bee" comes out from the head of the ox (*Geoponica* 15.2.30). This "ritual" of ox burying still existed in Egypt up to the 19th c.[317]. A useful clue for the dating of this belief could be that the Egyptian sign of the *bityw*, the honey-hunters (collectors of wild bees' honey) of Ramses III, was a bucranium[318]. A Neolithic origin from Anatolia however is not improbable. It is indeed likely that the oldest known traces of the belief come from the Neolithic settlement of Çatal Hüyük (7-6 millenium BCE) where we find depictions of bull heads in combination with bees, combs and double axe-like shapes[319]. Moreover the existence of such resemblances between the culture of Çatal Hüyük and that

---

[314] Jacob 1568, Wood 1870, Cook 1895, Shipley 1918, Leach 1949, 130, Davies 1986, 51. Kostas Tatsis, an apiarist from Epirus, ensured us that he has witnessed several beehives in donkeys' carcases what he called "a usual phenomenon in old times". In Lesbos the word "kouvani" means beehive but also carcase of a big animal (Makris 2000, 54)

[315] It was Nickel Jacob in 1568 who first made the correlation of the story of Samson with bougonia (reported by Fraser 1951, 12)

[316] Mago, from Carthage (c. 250 BCE) has been referred to as the father of Agriculture based on a 28 volume agricultural treatise written in Punic that is now lost. When Carthage was conquered by Rome (146 BCE) the Roman senate decreed that his agricultural works be translated. So it is possible that Virgil was familiar with the work of Mago. Vollgraff (1909, 42) believes that Virgil, in his fourth book of *Georgics*, followed the work of Nicander. This however, does not exclude that Virgil followed Mago too, especially on the subject of bougonia

[317] Cook 1965, 508-10 who gives however different interpretation of the custom

[318] Montet 1950, 24-25

[319] "shrine" VI.B.8 (Mellaart 1967, 112, 161-2, plates 40, 42)

of Minoan Crete prompted certain scholars to support the idea that the first Neolithic settlers of Crete came from Çatal Hüyük[320].

Another important question is the timing of bougonia, i.e. the morning rising of the Dog Star. This coincides with the time of honey harvesting. Pliny (*HN* 11.12) says that honey falls in the form of droplets from the sky[321] when the Dog Star rises. Aristotle (*Hist. an.* 553b) reports that honey is a summer product. Mead, which according to tradition was discovered by Aristaios, was also produced during the same season (Plin. *HN* 14.20). This period is up today the season of honey harvesting in most parts of Greece especially in the Aegean[322].

---

[320] Schachermeyr 1964, Dietrich 1974, 94-109

[321] The idea of the celestial origin of honey was widespread in antiquity (Antiphilos *Anth. Pal.*9.404; Phot. *Bibl.* 529b). Aristotle (*Hist.an.* 553b) claims that honey falls from the sky after the rise of the Pleiades constellation or when "*iris*" appears. Iris most probably refers to the coloured arc formed round the full moon (for its formation see Apul. *De Mundo* 16.14; Plin. *HN* 2.98, 2.150; Sen. *Q Nat.* 1.2, 1.3), since honey harvesting was performed mainly during full moon (Plin. *HN* 11.15). Even today in Greece traditional beekeeprs avoid harvesting honey but in full moon (Vrontis 1939, 205; Karalis personal communication). Theophrastus (fr. 190), Virgil (*G.* 4.1), Columella (*Rust.* 9.14.20) and Seneca (*Ep.* 84.4 and 85) also believed in the celestial origin of honey in the form of droplets. Perhaps the *Meliai* nymphs of Crete who according to Hesychius were bees and daughters of Ouranos (sky) reflect this belief for honey emanating from sky, the "*sky-honey*" (*αερομέλι*) as called by Amyntas (ap. Ath. 11.102). *Meliai* are probably mentioned in the Linear B tablets (KN Dv 1255, cf. Chadwick 1973, 560). The belief in the celestial origin of honey was widespread in many people (see Ransome 1937, 134 for examples from the Indian Vedas, the German and Finnish tradition and the Bible). See also for the same belief in Anatolia, Neufeld 1978 and Chouliara - Raios 1989, 31, n 48. In Byzantine times, Basilius (*Homiliae in hexaemeron* 8.4.30) says that honey falls from the sky "*νοτίς ενεσπαρμένην δροσοειδώς τοις άνθεσιν*" which recalls the "*ύεται η Ινδών γή δια του ήρος μέλιτι υγρώ* [...] *ένθα και μάλλον η δρόσος η γλυκεία κάθηται πεσούσα*" of Aelianus (*NA* 15.7). Even in 1609, Charles Butler (*The feminine monarchie*. London reported by Crane 2000a, 576) considered honey as distillation of air that falls as small flecks on the trees and the plants. The fact that honey is derived from the nectar of flowers after being processed by fermenting agents residing in liquids in the mouth of worker bees was realised only in the 18th c. (Crane 2000a, 577)

[322] The honey harvest of summertime refers to honey produced from thyme (it blooms during June, July and August), a plant which was (and is) considered the best for honey (Theophr. *Hist. pl.* 6.2.3; Varro *Rust.* 3.16.14; Palladius *Opus agriculturae* 2.37). Aristotle (*Hist. an* 554a) says that honey is gathered when the first wild figs ("*ερινεοί*") appear, that is to say in the beginning of August. There are of course other seasons too for honey harvesting. The harvest of autumn correspond to honey from autumnal heath (*Erica vertillata, Arbutus Uned*), ivy and pine while that of spring mainly refers to honey from fruit trees but also bay-laurel, asphodel and oak (for bee-plants and the period of their florescence see Nikolaidis 2000, 71 and *L' Abeille de France* 857, 154). Thus can be explained the three seasons of vintage of honey that are reported by the ancient writers. Aristotle apart from summertime reports vintage of honey during spring and also in autumn (*Hist. an* 626b). Varro (*Rust.* 3.16.34) reports the vintage at the rising of Pleiades (beginnings of May), the vintage in the end of summer (beginnings of September) and a third one, in the beginning of winter (beginnings of November). Virgil (*G* 4.231-235) reports two vintages of honey: one in May and one in November. Palladius (*Opus agriculturae* 7.7.1, 11.13.1) reports vintage of honey in June and October. In *Geoponica* (15.5.1), three seasons are reported: May, late autumn and October. In ancient Egypt two vintages of honey are reported (*Papiri greci e latini* 426.15)

Were the two activities of honey harvest and bougonia unrelated? We believe that the opposite is true, and the connection between the two can be established by the fact that in primitive apiculture the vintage of honey was a destructive process. It involved either the destruction of bees by fire – Aristaios was the first to smoke his bees with fire - or by poisoning with sulphur smoke[323], or the destruction of combs containing brood along with the ones containing honey which also often resulted in the death of the bees[324]. The collection of large quantities of honey can also have the same results since what remains does not suffice for the feeding and the survival of bees[325], thus leading to "pillage" (i.e. the raid of bees on foreign beehives in search of food). Virgil says that the bees of Aristaios died of "*sickness and hunger*". An additional destructive factor is the appearance of red wasps[326] and wax moths at this time of the year, both being significant bee enemies. These factors along with the famine of bees can lead the hive to an abnormal immigration (escape)[327]. Thus, during this period, there was a great need for producing new bees, a need which was fulfilled through bougonia. Sacrificing a bull to produce new bees – actually attracting other nearby swarms or impeding the escape of swarms - presupposes that honey was at least as valuable as oxen. The French dictum "*un essaim du mois May, vaut une vache du pays Bray*" and the corresponding English one "*a swarm of bees in May, is worth a cow and calf that day*"[328] are indicative of the swarm value. The Edict of Rothair in the *Lombardic Laws* imposed a fine of twelve solidi for stealing another man's hive. The importance of the offence can be deduced by the fact that no other kind of theft was punished more severely in the Edict[329]. In Anglo-Saxon England, also, the fines for bee-theft were quite high. Alfred reformed the laws, making fines for theft consistent, but he explicitly states that bee-theft was originally a crime with greater penalty than others, only comparable to the theft of gold[330]. A swarm of bees according to the Hittites' laws was worth the same as a sheep[331] and for Hesiod (*Op.* 232– 4) bees feature alongside sheep as the most important features of the hillsides.

We believe that bougonia, originally did not represent a religious ritual but a very primitive habit of beekeepers just like the "the attempt to rape Eurydice" does not represent a moral myth but the imposition of sexual abstinence before honey harvest.

---

[323] This practice is testified for the Byzantine period but also for many European countries, up to the beginning of the 20th c. (Liakos 1999a, 118) , Loukopoulos 1983, 397-8

[324] This is the case with skeps and other beehives without top movable bars. Destruction of the hive during honey harvesting was the rule in Arachova, Boeotia (*Kentro Laographias Acadimias Athinon*, 1938, mns 1153A, 19), in Pontos (Topalidis 1968-1969), in Florina (Bikos 1997b), in Andros (Bikos 1996c), in Sardinia (Crane 2000a, 188) and most parts of Europe, before the introduction of modern beehives, as well as in Cameroon, Africa (Crane 2000a, 53)

[324] Reported by Ransome 1937, 230

[325] See the strict instructions for limited quantities of honey harvesting in Varro (*Rust.* 3.16. 33), Columella (*Rust.* 9.15.8), Pliny (*HN* 11.35, 11.40) and *Geoponica* (15.5)

[326] Arist. *Hist. an.* 626a; Ael. *NA* 1.58. The wasps appear in the beginning of June (Nikolaidis 2000, 102) and especially in periods of drought (Loukopoulos 1983, 396)

[327] Varro *Rust.* 3.16. 33; Nikolaidis 2000, 39

[328] Reported by Ransome 1937, 230

[329] *Edict of Rothair*, sections 318-9 in Drew 1973, 114

[330] Libermann 1903, 54-5

[331] Goetze 1955, 193; Crane 2000a, 173

The two accounts by Virgil represent two prehistoric - probably of Egyptian origin - beekeeping practices. It is possible that Virgil drew these two accounts from the agriculture treatise of Mago who in turn was informed of these practices by traditional beekeepers who - just like modern apiarists - passed the secrets and the know-how of beekeeping from father to son in the form of a myth for many generations.

Works Cited

Aikaterinidis G.N. 1996. "κόμπος"; "κουδούνι" in *Εγκυκλοπαίδεια Πάπυρος Λαρούς Μπριτάννικα*. Vol. 35. Athens

Allaby M. ed. 1998. *A Dictionary of Ecology*. Oxford University Press. Oxford Reference Online (www.oxfordreference.com) Oxford University Press. 23 September 2005

Alp S. 1968. *Zylinder und Stempelsiegel aus Karahoyuk bei Konya*. Turk Tarih Kurumu Basimevi. Anakara.

Alphandery E. 1911. *L'apiculture par l'image*. Paris

Anagnostakis E. 2000. "Βυζαντινή μελωνυμία και μελίκρατος πότος. Αντιλήψεις για την χρήση των μελισσοκομικών προϊόντων στο Βυζάντιο ως τον 11° αι." in *Η Μέλισσα και τα προϊόντα της*. Πολιτιστικό Τεχνολογικό Ίδρυμα ΕΤΒΑ, Athens, 161-189

Anagnostopoulos I.T. 2000. "Δύο τύποι εγχώριων κυψελών της Φλωρινιώτικης μελισσοκομίας" in *Η Μέλισσα και τα προϊόντα της*. Πολιτιστικό Τεχνολογικό Ίδρυμα ΕΤΒΑ, Athens, 295-310

Anderson-Stojanovic V.R., Jones J.E. 2002. "Ancient beehives from Isthmia" *Hesperia* 71:345-376

Androudis P. 2000. "Το μέλι και το κερί στη μοναστηριακή ζωή των βυζαντινών χρόνων" in *Η μέλισσα και τα προϊόντα της*. Πολιτιστικό Τεχνολογικό Ίδρυμα ΕΤΒΑ. Athens. 211-220

Armbruster L. 1926. "Der Bienenstand als völkerkundliches Denkmal" *Bucherei Bienenk.* 8, 1-147

Armbruster L. 1928. "Die alte Bienenzucht der Alpen". *Bucherei Bienenk.* 9 :1-184

Barone R. 1984. *Anatomie comparée des mammifères domestiques*. Tome 3.Paris

Bass G.F. 1990. "A Bronze-Age writing diptych from the sea of Lycia" *Kadmos* 29:168-9

Baumbach L. 1968. *Studies in Mycenaean inscriptions and dialect 1953-1964*. Edizioni dell'Ateneo. Roma

Beinlich H., Hoffmann F. 1994. "Beinlich Ägyptische Wortliste" Göttinger Miszellen140,101-3. http://www.newton.cam.ac.uk/egypt/beinlich/beinlich.html#manuel (29/04/2005)

Betts A. D. 1922. "An Old Bee-Charm" *Bee World* 4. 140

Bikos T. 1994. "Μελισσο-αναφορές" *Μελισσοκομική Επιθεώρηση* 8(11):396-400

Bikos T. 1996. "Η μελένια κλεψύδρα" *Μελισσοκομική Επιθεώρηση* 10(3):113-7

Bikos T. 1997a. "Το σταθερό βήμα της μελισσοκομίας" *Μελισσοκομική Επιθεώρηση* 11(5):226-232

Bikos T. 1997b. "Μέλισσες από μάρμαρο" *Μελισσοκομική Επιθεώρηση* 11(11):485-488

Bikos T. 1997c. "Σκόρπιες πινελιές" *Μελισσοκομική Επιθεώρηση* 11(2):73-78

Bikos T. 1998. "Η κίνηση της κινητής" *Μελισσοκομική Επιθεώρηση* 12(12):536-541

Bikos T. 1999. "Στο νησί του Αρισταίου" *Μελισσοκομική Επιθεώρηση* 13(1):5-8

Bikos T. 2000. "Κινητή κηρύθρα" in *Η μέλισσα και τα προϊόντα της*. Πολιτιστικό Τεχνολογικό Ίδρυμα ΕΤΒΑ. Athens. 284-288

Black J., George A, Postgate N. (eds). 2000. *A Concise Dictionary of Akkadian*. Verlag

Blackman A.M. 1910. "Some Egyptian and Nubian notes" *Man*, 11

Bleek W. H., Lloyd L. C. 1911. *Bushman Folklore*. London

Bonet H.R., Mata C.P. 1997. "The archaeology of beekeeping in pre-Roman Iberia" *Journal of Mediterranean Archaeology* 10: 33-47

Breasted J.H. 1962. *Ancient records of Egypt*. Russell & Russell. New York

Buchholz H.G., Johrens G., Maull I. 1973. "Jagd und Fischfang: mit einem Anhang: Honiggewinnung" in *Archaeologia Homerica*. Bd.2 Vandenhoeck & Ruprecht. Göttingen

Buondelmonti C.1417. *Descriptio Insule Crete, liber Insularum*. ed Marie Anne van Spitael, Herakleia. Crete. 1981

Burke B. 1997. "The organization of textile production in Bronze Age Crete" in eds R. Laffineur, Ph. Betancourt "TEXNH. Craftsmen, Craftswomen and Craftsmanship in the Aegean Bronze Age / Artisanat et artisans en Égée à l'âge du Bronze." *Aegaeum* 16, 413-425

Burket W. 1987. *Homo necans: The Anthropology of Ancient Greek Sacrificial Ritual and Myth*. University of California Press. Berkeley

Cain C.D. 2001. "Dancing in the dark: Deconstructing a narrative of epiphany on the Isopata Ring" *AJA* 105(1):27-49

Callaghan P.J. 1992. "Archaic to Hellenistic pottery" in Knossos from Greek city to Roman colony. Excavations at the Unexplored Mansion. II. *British School at Athens*. Oxford. 89-480

Catling H.W. 1972. "An Early Byzantine pottery factory at Dhorios in Cyprus" *Levant* 4:1-82

Catling H.W., Catling E.A., Callaghan P., Smyth D. 1981. "Knossos 1975: Minoan paralipomena and post-Minoan remains" *BSA*:83-107

Chadwick J. 1973. *Documents in Mycenaean Greek*. 2n ed. Cambridge University Press. Cambridge

Chadwick J. 1976. *The Mycenaean world*. Cambridge University Press. Cambridge

Chantraine P. 1968. *Dictionnaire Etymologique de la Langue Grecque. Histoire des Mots*. Klincksieck. Paris

Chouliara-Raios E. 1989. *L'abeille et le miel en Egypte d'après les papyrus grecs*. Δωδώνη Παράρτημα No. 30. University of Ioannina. Ioannina

Chrisoulaki A. 1958. "Έθιμα των Σφακίων της Κρήτης" *Λαογραφία* 17(2):382-404

Christides V. 1984. *The conquest of Crete by the Arabs (ca 824). A turning point in the struggle between Byzantium and Islam*. Akademia Athenon. Athens

Clark G. 1942. "Bees in antiquity" *Antiquity* 16:208-215

Contenau G. 1937. *La civilisation d'Assur et de Babylone*. Payot. Paris

Cook A.B. 1895. "The bee in Greek mythology" *JHS* 15:18

Cook A.B. 1964-1965. *Zeus. A study in ancient religion*. Vol. I (1964), Vol. II (1965). Biblo and Tannen. New York

Cotton C. 1842. *My bee book*. London

Cowan F. 1865. *Curious facts in the history of insects*. Philadelphia

Crane E. 2000a. *The world history of beekeeping and honey hunting*. (second impression) Duckworth. London

Crane E. 2000b. "The Transmission of Beekeeping Round the Ancient Mediterranean" abstract from *Bee-keeping in the Graeco-Roman World, a conference organised by Simon Price and Lucia Nixon at Lady Margaret Hall*, Oxford, on 7 November 2000

Crane E., Walker P. 2000c. "Wall recesses for bee hives" *Antiquity* 286:805-11

D'Agata A.L. 1992. "Late Minoan Crete and horns of consecration: a symbol in action" in Laffineur R., Crowley J.L (eds) EIKΩN Aegean Bronze Age Iconography: Shaping a Methodology. *Aegaeum* 8. Universite de Liege, 247-255

Danielidou D. 1998. *Η οκτώσχημη ασπίδα στο Αιγαίο της 2ης π.Χ χιλιετίας*. Ακαδημία Αθηνών. Σειρά Μονογραφιών 5. Athens

Davaras C. 1984. "Μινωικό κηριοφόρο πλοιάριο της Συλλογής Μητσοτάκη" *Αρχ. Εφ.*, 55-93

Davaras C. 1986. "A new interpretation of the ideogram *168" *Kadmos* 25, 38-43

Davaras C. 1989. "Μινωικά μελισσουργικά σκεύη". *Αρχαιολογική Εταιρεία Αθηνών, Φίλια έπη εις Γεώργιον Ε. Μυλωνάν* Vol. 3. Βιβλιοθήκη της εν Αθήναις Αρχαιολογικής Εταιρείας 103.Athens. 1-7

Davaras C. 1992. *Μινωικός και Ελληνικός Πολιτισμός από τη Συλλογή Μητσοτάκη*, Edit. Maragou L. Ίδρυμα Ν. Γουλανδρή, Μουσείο Κυκλαδικής Τέχνης, Athens, p.106

Dave K.N. 1954/55. "Beekeeping in Ancient India" *Indian Bee J.* 16,17: article in 10 parts

Davies N. de G. 1944. *The tomb of Rekhmire at Thebes*. N.H. Ayer Co. Salem

Debeauvoys C.P. 1846. *Guide de l'apiculteur*. 5th ed. 1856. Cosnier et Lachese. Angers

Decavalla O. 2007. "Beeswax in Neolithic perforated sherds from the northern Aegean: new economic and functional implications" in Mee C. and Renard J. (eds) *Cooking Up the Past: Food and Culinary Practices in the Neolithic and Bronze Age Aegean*. Oxford, Oxbow, p. 148ff

DeJesus P.S. 1980. *The development of prehistoric mining and metallurgy in Anatolia*. BAR Inter. Ser. 74. Oxford

Demetropoulos A., Ioannidis I. 2002. *Ερπετά της Ελλάδας και της Κύπρου*. Μουσείο Γουλανδρή Φυσικής Ιστορίας. Athens.

Demopoulou N. 2005. Το αρχαιολογικό μουσείο Ηρακλείου. Eurobank. Athens

Demopoulou N., Rethemiotakis G. 2000. "The sacred conversation ring from Poros" in *Corpus der Minoischen und Mykenischen Siegel*. Beiheft 6: Minoisch-Mykenische Glyptik. Stil, Ikonographie, Funktion. V. Internationales Siegel-Symposium. Marburg, 23.-25. September 1999. Ed. Müller W. Mann Verlag. Berlin. 39-56

Di Vita A. 1993. *Annuario della Scuola Archeologica di Atene*. 66/67

Dietrich B.C. 1996. "religion, Minoan and Mycenaean" in *Oxford Classical Dictionary*. Oxford University Press. (eds). Hornblower S., Spawforth A. 3d ed. Oxford, New York

Drew Fisher K. (trans) 1973. *The Lombard Laws*. University of Philadelphia Press. Philadelphia

Duhoux Y. 1975. "Les Idéogrammes *168 et *181 du linéaire B" *Kadmos* 14 ;117-121

Duhoux Y. 1977. *Le disque de Phaestos. Archeologie, Epigraphie, Edition critique, Index*. Peeters. Louvain

Dumay R. 1997. *Le Rat et l'Abeille. Court traité de gastronomie préhistorique*. Phébus. Paris

Dundes, A. (ed.) 1981. *The Evil Eye: A Folklore Casebook*. Garland Publishing. New York

Eckert G. 1943. *Die Wanderbienenzucht in der Chalkidike*. Thessaloniki

Eco U. 1976. *A Theory of Semiotics*. Bloomington, IN: Indiana University Press/London: Macmillan

Efthimiou-Hatzilakou M. 1981-2. "Μικρό σημείωμα για το μελισσοκόφινο" *Εθνογραφικά* 3: 125-6

Elworthy F.T. 2004. *The evil eye*. Dover Publications. New York

Erman A. 1909. *Die Aegyptische Religion*. Druck und Verlag Georg Reimer. Berlin

Evans A. 1901 "Mycenaean Tree and Pillar Cult and its Mediterranean relations" *JHS* 21:99-204

Evans A. 1909. *Scripta Minoa*. Oxford

Evans A. 1921-1935. *The Palace of Minos*. 4 volumes. Biblo and Tannen. London

Evershed, R. P., Vaughan S.J., Dudd S.N., Soles J.S. 1997. "Fuel for thought? Beeswax in lamps and conical cups from late Minoan Crete" *Antiquity* 71: 979-985

Fahd T. 1968. "L'abeille en Islam" in Chauvin R. (ed) *Traité de biologie de l'abeille*. Masson. Paris, V, 61-83

Faure P. 1999. *Η καθημερινή ζωή στην Κρήτη τη Μινωική εποχή*. Translated by E. Aggelou. Παπαδήμας. Athens

Fife A.E. 1964. "Christian swarm charms from the ninth to the nineteenth centuries." *Journal of American Folklore* 77(304):154-9

Filotheou G. 1980. "Λαογραφικά Λανίας. Μελισσοκομία" *Λαογραφική Κύπρος* 10(28-30): 81-83

Forbes R.J. 1966. *Studies in Ancient Technology*. v.5. E.J.Brill. Leiden

Foster KP. 1995. "A flight of swallows" *AJA* 99:409-425

Francis J. 2000. "Finds of Graeco-Roman Beehives from Sphakia, SW Crete" abstract from *Bee-keeping in the Graeco-Roman World, a conference organised by Simon Price and Lucia Nixon at Lady Margaret Hall*, Oxford, on 7 November 2000

Francis J. 2001. "Beehives and Beekeeping in Ancient Sphakia". *9th International Congress of Cretan Studies*. 1-6 October. Elounda

Fraser H.M. 1951. *Beekeeping in antiquity*. University of London Press. London

Frazer G. 1936. *The Golden Bough*. 3rd ed. Vol. 3. Macmillan. London.

Frier B.W. 1982/1983. "Bees and Lawyers" *CJ* 78(2):105-14

Gardiner H. 1957. *Egyptian Grammar*. 3d ed. Griffith Institute. Oxford

Gennadios P.G. 1914. *Λεξικόν Φυτολογικόν*. Τροχαλία. Athens

Geny-Tsofopoulou E. 2002. "Πήλινη Κυψέλη" in Papanikola-Bakirtze D. (ed) *Καθημερινή ζωή στο Βυζάντιο*. Υπουργείο Πολιτισμού. Athens 2002

Georgandas D. 1957. "The forerunner of the modern hive" *Bee World* 38(11): 286-9

Gianopoulou M., Demesticha S. 1998. *Τσακαλαριά*. Κέντρο Μελέτης Νεώτερης Κεραμικής Κοινότητας Μανταμάδου. Athens

Giousaris A. 2000. "Παραδοσιακή μελισσοκομία στην Καρδίτσα πριν το 1960. Πλίνθινοι τοίχοι αντί κυψελών" in *Η Μέλισσα και τα προϊόντα της*. Πολιτιστικό Τεχνολογικό Ίδρυμα ΕΤΒΑ, Athens, 311-323

Glotz G. 1923. *La civilisation Egeene*. Editions de la Renaissance du Livre. L'évolution de l'humanité N° 9. Paris

Goetze A. 1955. "The Hittite Laws" in *Ancient Near Eastern texts relating to the Old Testament*. Prichard J.B. ed. 2nd ed. Princeton University Press. New Jersey, 188-197

Gombrich E.H. 1977. *Art and Illusion: A Study in the Psychology of Pictorial Representation*. Phaidon. London

Gow A.S.F. 1944. "Βουγονία in Geoponica XV.2" *The Classical Review* 58, 14-15

Graham A.J. 1975. "Beehives from Ancient Greece" *Bee World* 56:64-75

Grumach E. 1964. "Ein kretisches Honigzeichen?" *KretChron* 18, 7-14

Haldane J.B.S. 1955. "Aristotle's account of bees' dances" *JHS* 75, 24-25

Hallager E. 1985. *The Master Impression. SIMA* LXIX

Harrison J.E. 1903. *Prolegomena to the study of Greek Religion*. Reprint edition 1991. Princeton University Press, Princeton

Hatzaki-Kapsomenou C. 2001. *Θησαυρός νεοελληνικών αινιγμάτων*. Πανεπιστημιακές εκδόσεις Κρήτης. Heraklion.

Hayes J.W. 1983. "The Villa Dionysos excavations, Knossos: the pottery" *BSA* 78:97-170

Hogarth D.G. 1902. "The Zakro sealings" *JHS* 22:76-93

Homann-Wedeking B. 1950. "A kiln site at Knossos" *BSA* 45:165-192

Hood S. 1976. "The Mallia Gold Pendant: Wasps or Bees?" In Emmison F.G. and R. Stephens (eds). *Tribute to an Antiquary. Essays Presented to Marc Fitch by Some of His Friends*. Leopard's Head Press, London. 59-72

Horn T. 2005. *Bees in America: How the Honey Bee Shaped a Nation*. University Press of Kentucky, Lexington.

Hornell J. 1925. "Horns in Madeiran Superstition". *The Journal of the Royal Anthropological Institute of Great Britain and Ireland*. 55:303-310

Houlihan, P E 1986 *The Birds of Ancient Egypt*. Aris and Phillips. Warminster

Hudson-Williams T. 1935. "King Bees and Queen Bees" *The Classical Review* 49:2-4

Ibn Al-'Awwam. 2000. *Le livre de l'agriculture*. Thesaurus. Actes Sud/Sindbad

Ifantidis M.D. 1983. "The movable-nest hive: a possible forerunner to the movable-comb hive" *Bee World* 64(2):79-87

Janko R. 1982. *Homer, Hesiod and the Hymns*. Cambridge University Press. Cambridge

Jones B. 2001. "The Minoan Snake Goddess. New Interpretations of her Costume and Identity" in Laffineur R., Hägg R. (eds). "POTNIA. Deities and Religion in the Aegean Bronze Age". *Aegaeum* 22: 259-264. Proceedings of the 8th International Aegean Conference Göteborg, Göteborg University, 12-15 April 2000

Jones J.E. 1976. "Hives and Honey of Hymettus: beekeeping in Ancient Greece" *Archaeology* 29.2:80-91

Jones J.E. 2000. "Hives from Isthmia and Elsewhere" abstract from *Bee-keeping in the Graeco-Roman World, a conference organized by Simon Price and Lucia Nixon at Lady Margaret Hall*, Oxford, on 7 November 2000

Jones J.E., Graham A.J., Sackett L.H. 1973. "An Attic Country House below the Cave of Pan at Vari" *Annual of the British School at Athens* 68:355-452

Jung H. 1989. "Methodisches zur hermeneutic der Minoischen und Mykenischen bilddenkmaler" in *Corpus der Minoischen und Mykenischen Siegel*. Beiheft 3: Fragen und Probleme der bronzezeitlichen ägäischen Glyptik. Beiträge zum 3. Internationalen. Marburger Siegel-Symposium 5.- 7. September 1985. Ed. W. Müller. Mann Verlag. Berlin, 91-109

Kaplan M. 1992. *Les hommes et la terre a Byzance du Ve au XIe siecle*. Publications de la Sorbonne. Byzantina Sorbonensia 10. Paris

Katsouleas S.G. 2000. "Ο σχετικός με την μέλισσα γλωσσικός πλούτος. Ο όρος κυψέλη" in *Η Μέλισσα και τα προϊόντα της*. Πολιτιστικό Τεχνολογικό Ίδρυμα ΕΤΒΑ, Athens, 339-370

Kenna V.E.G. (ed). 1967. *Corpus der Minoischen und Mykenischen Siegel*. Bd. VII: Die englischen Museen II: London, British Museum - Cambridge, Fritzwilliam Museum - Manchester, University Museum - Liverpool, City Museum - Birmingham, City Museum. Mann Verlag. Berlin

Kenna V.E.G. (ed). 1972. *Corpus der Minoischen und Mykenischen Siegel*. Bd. XII: Nordamerika I: New York, The Metropolitan Museum of Art. Mann Verlag. Berlin

Kilian-Dirlmeier I. 1987"Das Kupperlgrab von Vapheio" *JRGZM* 34:197-212

Kitchell K.F. 1981. "The Mallia "wasp" pendant reconsidered" *Antiquity* 55, 9-15

Konstantinidis K. 2000. "Εξέλιξη της παραδοσιακής κυπριακής κυψέλης" in *Η Μέλισσα και τα προϊόντα της*. Πολιτιστικό Τεχνολογικό Ίδρυμα ΕΤΒΑ, Athens, 324-29

Konstantopoulos C.G. 1987. *Η μαθητεία στις κομπανίες των χτιστών της Πελοποννήσου*. Ιστορικό Αρχείο Ελληνικής Νεολαίας. Athens

Korre-Zografou K. 2008. *Άνθρωποι και παραδοσιακά επαγγέλματα στο Αιγαίο. II.* Ίδρυμα Μείζονος Ελληνισμού. Athens

Kostakis T. P. 1963. *Η Ανακού*. Κέντρο Μικρασιατικών Σπουδών. Athens

Koutri S. 1999. *Κεραμικές μορφές της Λέσβου*. Eommex, Ίνδικτος. Athens

Kueny G. 1950. "Scènes apicoles dans l'ancienne Egypte" *JNEStud* 9,84-93

Kukules Ph. 1948. *Βυζαντινών βίος και πολιτισμός*. Volume ΑΙΙ. Γαλλικό Ινστιτούτο Αθηνών. Athens

Kukules Ph. 1951. "Η μελισσοκομία παρά Βυζαντινοίς" *Byzantinische Zeitschrift* 44:347-357

Kyriakidis E. 2005. "Unidentified Floating Objects on Minoan Seals" *AJA* 109,137-54

Kyrou D. 2000. "Η μελισσοκομία στην οικονομία και τον καθημερινό βίο της Αρναίας σε παλαιότερες εποχές" in *Η Μέλισσα και τα προϊόντα της*. Πολιτιστικό Τεχνολογικό Ίδρυμα ΕΤΒΑ, Athens, 371-89

Laffineur R. 1990. "The Iconography of Mycenaean Seals and the status of their owners" *Aegaeum* 6, 117-160

Laffineur R. 1992. "Iconography as evidence of social and political status in the Mycenaean Greece" in Laffineur R., Crowley J.L (edts) "ΕΙΚΩΝ. Aegean Bronze Age Iconography: Shaping a methodology". *Aegaeum* 8. Universite de Liege. 105-111

Langdon M.K. 1985. "Hymettiana I". *Hesperia*, 54(3): 257-270

Lapatin K. 2003. *Mysteries of the Snake Goddess. Art, desire and the forging of history*. Da Capo Press.

Larson J. 1995. "The Corycian Nymphs and the Homeric Hymn to Hermes" *GRBS* 86:341-57

Lawall M.L., Papadopoulos J.K., Lynch K.M., Tsakirgis B., Rotroff S.I., MacKay C. 2001. "Notes from the Tins: Research in the Stoa of Attalos, Summer 1999" *Hesperia* 70: 163-182.

Lawson J.C. 1964. *Modern Greek folklore and ancient Greek religion*. University Books. New York

Leake, W. 1830. *Travels in the Morea with a Map and Plans*. 3 Vols. London

Leclant J. 1968. "L'abeille et le miel dans l'Egypte pharaonique" in Chauvin R. (ed). *Traité de biologie de l'abeille*. Masson. Paris, V, 51-60

Lefebvre G. 1929. *Histoire des grands prêtres d'Amon de Karnak*. Paris

Leontidis T. 1986. *Τα Κρητικά καλάθια. Μορφολογική, κατασκευαστική μελέτη*. Μουσείο Κρητικής Εθνολογίας. Athens

Lexikon der Agyptologie. 1975. Helck W., Otto E. (eds). Otto Harrassowitz. Wiesbaden

Liakos B. 1996. "Το κερί" *Μελισσοκομική Επιθεώρηση* 10(10): 370-4

Liakos B. 1999a. "Η μελισσοκομία στον Ελλαδικό χώρο, κατά τους Βυζαντινούς και Μεταβυζαντινούς χρόνους και η συμβολή της στην ανάπτυξη της σύγχρονης μελισσοκομίας" *Μελισσοκομική Επιθεώρηση* 13(3):115-8

Liakos B. 1999b. "Σε ποιόν ανήκει ο αφεσμός σύμφωνα με τις Ασσίζες της Κύπρου" *Μελισσοκομική Επιθεώρηση* 13(7-8): 300-1

Libermann F. 1903. *Gesetze der Angeelsachsen*. Max Niemeyer. Halle am Saale

Liddell H.G., Scott R., 1940. *Greek-English Lexicon*, 9th edn. Rev. H. S. Jones Clarendon Press. Oxford

Lombard M. 1812. *Manuel des propriétaires des abeilles*. Paris

Lombard M. 1994. *L' Islam dans sa première grandeur (VIIIe - XIe siecle)*. Flammarion. Paris

Loukatos D.S. 1992. *Εισαγωγή στην Ελληνική Λαογραφία*. 4th ed. MIET. Athens

Loukopoulos D. 1983. *Γεωργικά της Ρούμελης*. Δωδώνη. Athens

Lucas A., Harris J.R. 1999. *Ancient Egyptian materials and industries*. 4th edn. Dover Publications. New York

Ludorf G. 1998/1999. "Leitformen der attichen Gebrauchskeramik: Der Bienenkorb" *Boreas* 21-22, p.41-130

Lykiardopoulos A. 1981. "The evil eye: Towards an exhaustive study" *Folklore* 92(2):221-30

Makris T. 2000. *Πολιχνιάτικα*. Municipality of Lesbos. Athens

Marchenay P. 1979. *L'Homme et l'abeille*. Berger-Levrault. Paris

Marconi M. 1940. "Μέλισσα, dea Cretesa" *Athenaum* 164-168

Marinatos N. 1989. "The tree as a focus of ritual action in Minoan glyptic art" in *Corpus der Minoischen und Mykenischen Siegel*. Beiheft 3: Fragen und Probleme der bronzezeitlichen ägäischen Glyptik. Beiträge zum 3. Internationalen. Marburger Siegel-Symposium 5.- 7. September 1985. Ed. W. Müller. Mann Verlag. Berlin. 127-142

Martin G. T. 1971. *Egyptian Administrative and Private Name Seals Principally of the Middle Kingdom and Second Intermediate Period*. Oxford University. London

Mazar A., Panitz-Cohen N. 2007. "It is the land of honey: Beekeeping in Tel Rahov" *Near Eastern Archaeology* 70(4):202-219

Megas G. 1941. "Ζητήματα Ελληνικής Λαογραφίας. 2". *Επετηρίς του Λαογραφικού Αρχείου της Ακαδημίας Αθηνών*. Athens

Meier G. 1941/44. "Die zweite Tafel der Serie bit meseri" *AOF* 14, 139-152

Melas M. 1999. "The Ethnography of Minoan and Mycenaean Beekeeping" in Betancourt Philip P., Vassos Karageorghis, Robert Laffineur, and Wolf-Dietrich Niemeier. (eds). Meletemata: Studies in Aegean Archaeology Presented to Malcolm H. Wiener as He Enters His 65th Year. Vol. II. Université de Liege : Histoire de l'art et archéologie de la Grèce antique; University of Texas at Austin: Programs in Aegean Scripts and Prehistory. *Aegaeum* 20:485-491

Mendelsohn I. 1940. "Guilds in Ancient Palestine". *Bulletin of the American Schools of Oriental Research* 80:17-21

Michalopoulou-Veikou X.G. 1996. *Το μάτιασμα: Η κοινωνική δυναμική του βλέμματος σε μία κοινότητα της Μακεδονίας*. Dissertation. Πάντειο Πανεπιστήμιο Κοινωνικών και Πολιτικών Επιστημών. Τμήμα Κοινωνικής Πολιτικής και Κοινωνικής Ανθρωπολογίας. Athens

Militarev A., Kogan L. *Semitic Etymological Dictionary*. Vol. 1. Anatomy of man and animals. Ugarit-Verlag. Münster 2000

Montet P. 1950. "Etudes sur quelques prêtres et fonctionnaires du dieu Min" *JNES* 9:18-27

Moret A. 1908. *Au temps de pharaons*. Armand Colin. Paris

Morgan L. 1989. "Ambiguity and Interpretation in Corpus der Minoischen und Mykenischen Siegel" in *Corpus der Minoischen und Mykenischen Siegel*. Beiheft 3: Fragen und Probleme der bronzezeitlichen ägäischen Glyptik. Beiträge zum 3. Internationalen. Marburger Siegel-Symposium 5.- 7. September 1985. Ed. W. Müller. Mann Verlag. Berlin. 145-161

Morpurgo Davies A. 1979. "Terminology of Power and terminology of work in Greek and Linear B" in E. Rich, H. Muhlestein (eds), *Colloquium Mycenaeum*, 87-108

Morris S.P. 1992. "Prehistoric Iconography and historical sources: Hindsight through texts:" in Laffineur R., Crowley J.L (eds) ΕΙΚΩΝ Aegean Bronze Age Iconography: Shaping a Methodology. *Aegaeum* 8. Universite de Liege, 209-215

Morris S.P. 2001. "The prehistoric background of Artemis Ephesia: a solution to the enigma of her breasts?" in Muss U. ed. *Der Kosmos der Artemis von Ephesos*. Sonderschriften des Österreichischen Archäologischen Instituts, Band 37, Wien. 135-151

Müller W., Pini I. (eds). 1998. *Corpus dre Minoischen und Mykenischen Siegel*. Bd. II: Iraklion, Archäologisches Museum. Teillieferung 7: Die Siegelabdrücke von Kato Zakros unter Einbeziehung von Funden aus anderen Museen. Nach Vorarbeiten von Nikolaos Platon. Mann Verlag. Berlin

Müller W., Pini I. (eds). 1999. *Corpus de Minoischen und Mykenischen Siegel*. Bd. II: Iraklion, Archäologisches Museum. Teillieferung 6: Die Siegelabdrücke von Agia Triada und anderen zentral- und ostkretischen Fundorten unter Einbeziehung von Funden aus anderen Museen. Nach Vorarbeiten von Nikolaos Platon. Mann Verlag. Berlin

Mylonas G.E. 1966. *Mycenae and the Mycenaean Age*. PUP. Princeton

Mynors R.A.B. 1994. *Virgil. Georgics*. Clarendon. Oxford

Neustadt E. 1906. *De Iove Cretico*. Berlin

Newberry P.E. 1938. "Bee-hives in Upper Egypt" *Man* 38:31-32

Nikolaidis N. 1955. "Facts about beekeeping in Greece" *Bee World* 36(8): 141-149

Nikolaidis N. 2000. *Η μελισσοκομία χωρίς δάσκαλο*. Athens

Nilsson M.P. 1950. *The Minoan Mycenaean Religion and its survival in Greek Religion*. 2nd edn. Biblo and Tannen. Lund

Nixon L. 2000. "Traditional beekeeping in Sphakia, SW Crete" abstract from *Bee-keeping in the Graeco-Roman World, a conference organized by Simon Price and Lucia Nixon at Lady Margaret Hall*, Oxford, on 7 November 2000

Noe S. 1958. *The Coinage of Kaulonia*. ANS Numismatic Studies No. 9

Nouaros M.G.M. 1934. *Λαογραφικά σύμμεικτα Καρπάθου*. Athens

Ober J. 1981. "Rock-Cut Inscriptions from Mt. Hymettos" *Hesperia*, 50(1):68-77

Oikonomidis D. 1965-1966. "Η μελισσοκομία εν Νάξο και εν Ανάφη" *Επετηρίς Εταιρείας Κυκλαδικών Μελετών* 5 : 617-634

Olivier JP., Godart L., Seydel C., Sourvinou C. 1973. *Index Généraux du Linéaire B*. Edizioni Dell'Ateneo. Roma

Onassoglou A. 1981. "Die kombinationen der talismanischen siegel". In *Corpus der Minoischen und Mykenischen Siegel*. Beiheft 1, (eds) Matz F., Pini I. Gebr. Mann Verlag. Berlin, 117-133

Pager H. 1971. *Ndedema*. Akademische Druck – u. Verlagsanstalt. Graz

Panaretos A. 1967. *Κυπριακή γεωργική λαογραφία*. Πρόοδος. Nicosia.

Papaefthimiou-Papanthimou A., Pilali-Papasteriou A. 1994. *ΑΕΜΘ*, 8 :80-90

Papaefthimiou-Papanthimou A., Pilali-Papasteriou A. 1997. *Οδοιπορικό στην Προϊστορική Μακεδονία*. Παρατηρητής. Thessaloniki

Papaefthimiou-Papanthimou A., Pilali-Papasteriou A. 1998. "Chroniques des fouilles et découvertes archéologiques en Grèce en 1996 et 1997" *BCH* 122:855

Papagelos I.A. 2000. "Η μελισσοκομία στην Χαλκιδική κατά τους μέσους χρόνους και την τουρκοκρατία" in *Η μέλισσα και τα προϊόντα της* Πολιτιστικό Τεχνολογικό Ίδρυμα ΕΤΒΑ. Athens. 190-210

Papapostolou I.A. 1977. *Τα σφραγίσματα των Χανίων*. Η εν Αθήναις Αρχαιολογική Εταιρεία. Athens

Papatsaroucha E. 2005. "La Pierre et l'objet double: questions iconographiques de la glyptique Minoenne" in Bradfer-Burdet I., Detournay B., Laffineur R. (eds) Κρης Τεχνίτης, L'Artisan Crétois, *Aegaeum* 26 :177-183, Université de Liege. University of Texas at Austin: Programs in Aegean Scripts and Prehistory.

Perles C. 2001. *The Early Neolithic in Greece*. Cambridge University Press. Cambridge

Persson A.W. 1942. *The religion of Greece in prehistoric times*. University of California Press. Berkeley

Petanidou T., Smets E. 1995. "The potential of marginal lands for bees and apiculture: Nectar secretion in Mediterranean shrublands". *Apidologie* 26: 39-52

Petropoulos D. 1957. "Μελισσοκομικά Χαλκιδικής και δυτικής Μακεδονίας" *Λαογραφία ΙΖ*, 186-196 and 356-357

Petropoulos M. "The Geometric Temple of Ano Mazaraki (Rakita) in Achaia during the Period of Colonization", in Emanuele Greco (a cura di), *Gli Achei e l'dentità etnica degli Achei d'Occidente*, (Tekmeria, 3) Paestum - Atene, 2002, pp. 143-164

Phiorou E. 2005. "Ο μελισσοκόμος" *Γεωτρόπιο* 275: 46-53. Athens

Picard C. 1940. "L'Ephesia, les Amazones et les abeilles" *REA* 42, 270-284

Piccirillo M. 1993. *The Mosaics of Jordan*. American Center of Oriental Research. Amman

Pini I. (ed). 1970. *Corpus der Minoischen und Mykenischen Siegel*. Bd. II: Iraklion, Archäologisches Museum. Teillieferung 5: Die Siegelabdrücke von Phästos. Mann Verlag. Berlin

Pini I., Betts J.H., Gill M.A.V., Sürenhagen D., Waetzoldt H. (eds). 1988. *Corpus der Minoischen und Mykenischen Siegel*. Bd. XI: Kleinere Europäische Sammlungen. Mann Verlag. Berlin

Platakis E. 1977. "Ζωώνυμα σπήλαια της Κρήτης" *Κρητολογία* 4:123-132

Platon L. 2005. "Το φαράγγι των νεκρών" *Κρητικό Πανόραμα* 9, 45-57

Platon N. 1948. "Nouvelle interprétation des idoles cloches du minoen moyen I" *Revue archéologique* 6(31-32): 833-846

Platon N. 1971. *Zakros*. Scribner's Sons. New York

Platon N., Pini I. (eds). 1984. *Corpus de Minoischen und Mykenischen Siegel*. Bd. II: Iraklion, Archäologisches Museum. Teillieferung 3: Die Siegel der Neupalastzeit. Mann Verlag. Berlin

Polites N. 1975. *Λαογραφικά Σύμμεικτα Β*. Ακαδημία Αθηνών. Κέντρο Ερεύνης Ελληνικής Λαογραφίας. Athens

Polunin O. 1987. *Flowers of Greece and the Balkans*. Oxford University Press. Oxford and New York

Pritchett W.K. 1956. "The Attic Stelai" *Hesperia* 25:178-317

Protopsaltis G. 2000. "Ποροκοπείο κυψελών στα Κύθηρα" in *Η μέλισσα και τα προϊόντα της*. Πολιτιστικό Τεχνολογικό Ίδρυμα ΕΤΒΑ. Athens. 289-294

Psilakis N. 2005. *Λαϊκές τελετουργίες στην Κρήτη. Έθιμα στον κύκλο του χρόνου*. Καρμανώρ. Heracleion.

Rackman O., Moody J. 2004. *Η δημιουργία του Κρητικού τοπίου* (transl. Sbonias Κ.) Πανεπιστημιακές Εκδόσεις Κρήτης. Heracleion.

Rambach J. 2002. "Olympia. 2500 Jahre Vorgeschichte von der Gründung des eisenzeitlichen griechischen Heiligtums" in *Olympia 1875-2000. 125 Jahre Deutsche Ausgrabungen*. Berlin 9.-11. November 2000. Helmut Kyrieleis ed. Philipp von Zabern, 177-212

Rammou A., Bikos T. 2000. "Η Ελλάδα της μελισσοκομίας. Τρία χρόνια μελισσοκομικών καταγραφών" in *Η μέλισσα και τα προϊόντα της*. Πολιτιστικό Τεχνολογικό Ίδρυμα ΕΤΒΑ. Athens. 413-435

Ransome H.M. 1937. *The sacred Bee*. George Allen and Unwin. London

Rehak P., Younger J.G. 2001. "Review of Aegean Prehistory VII: Neopalatial, Final Palatial, and Postpalatial Crete" in Cullen T. (edt) Aegean Prehistory. A review. *AJA Supplement I*, Boston, 383-465

Renfrew C. 1972. *The Emergence of civilization. The Cyclades and the Aegean in the Third Millennium BC*. Methuen. London

Renfrew, C. 1999. "The Loom of Language and the Versailles Effect" In MELETEMATA: Studies in Aegean Archaeology Presented to Malcolm H.Wiener as he enters his 65th Year. Vol. 3, edited by Philip P. Betancourt, Vassos Karageorghis, Robert Laffineur and Wolf-Dietrich Niemeier, 711-20. *Aegaeum* 20. Liège: Université de Liège.

Reras I. 2001. *Σφήκες*. Μελισσοκομική Επιθεώρηση. Thessaloniki

Richards-Mantzoulinou E. 1979. "Μέλισσα Ποτνία" *Αρχαιολογικά Ανάλεκτα εξ Αθηνών* (*Athens Annals of Archaeology*) 12:72-89

Rizopoulou-Igoumenidou E. 2000. "Η παραδοσιακή μελισσοκομία στην Κύπρο και τα προϊόντα της (μέλι, κερί) κατά τους νεώτερους χρόνους" in *Η μέλισσα και τα προϊόντα της*. Πολιτιστικό Τεχνολογικό Ίδρυμα ΕΤΒΑ. Athens. 390-408

Rocca abbe della. 1790. *Traité complète sur les abeilles avec une nouvelle méthode de les gouverner telle qu'elle se pratique à Syra, précède d'un précis historique et économique de cette île*. 3 vols. Paris

Romaios K.A 1955. "Το κακό μάτι" *Μικρά Μελετήματα* 91-98. Thessaloniki

Rouse W.H.D. 1896. "Folklore firstfruits from Lesbos" *Folklore* 7(2):142-61

Rutkowski B. 1981. *Frühgriechische Kultdarstellungen*. Beiheft 8. Athenische Mitteilungen. Berlin

Rutkowski B. 1984. *The cult places of the Aegean*. New Haven. Yale University Press, 99-118

Sakellarakis J.A., Kenna V.E.G. (eds). 1969. *Corpus der Minoischen und Mykenischen Siegel*. Bd. IV: Iraklion, Sammlung Metaxas. Mann Verlag. Berlin

Sakellariou A. (ed). 1964. *Corpus de Minoischen und Mykenischen Siegel*. Bd. I: Die minoischen und mykenischen Siegel des Nationalmuseums in Athen. Mann Verlag. Berlin

Sakellariou A. 1966. *Μυκηναϊκή Σφραγιδογλυφία*. Athens

Sakellariou A. 1995. "Les bagues-cachets Creto-Myceniennes: Art et Fonction" in *Corpus der Minoischen und Mykenischen Siegel*. Beiheft 5: Sceaux Minoens and Myceniens. Ed. W. Müller. Mann Verlag. Berlin, 313-329

Sanders I.F. 1982. *Roman Crete. An Archaeological Survey and Gazetteer of Late Hellenistic, Roman, and Early Byzantine Crete*. Warminster. Wilts

Schliemann. H. 1878. *Mycenae*. London

Schoep I. 1999, 217 "Tablets and Territories? Reconstructing Late Minoan IB Political Geography through undeciphered Documents" *AJA* 103:201-21.

Schuchhardt C. 1891. *Schliemann's excavations*. trans. Sellers E. London

Shaw I. 2003. *The Oxford History of Ancient Egypt*. New edition. Oxford University Press. Oxford

Simon E. 1983. *Festivals of Attica. An archaeological commentary*. The University of Wisconsin Press. Wisconsin

Soles J.S. 1991. "The Gournia Palace" *AJA* 95:17-78

Somville P. 1978. "L'abeille et le taureau, ou la vie et la mort dans la Crete minoenne." *Revue de l'histoire des religions* 4 :129-146

Sourvinou-Inwood C. 1989. "Space in late Minoan religious scenes in glyptic – some remarks" in *Corpus der Minoischen und Mykenischen Siegel*. Beiheft 3: Fragen und Probleme der bronzezeitlichen ägäischen Glyptik. Beiträge zum 3. Internationalen. Marburger Siegel-Symposium 5.- 7. September 1985. Ed. W. Müller. Mann Verlag. Berlin, 241-257

Spamer J.B. 1978. "The old English bee charm: an explication" *Journal of Indo-European studies* 6:279-91

Swammerdam J. 1737. *Biblia Naturae*. Leyde

Tatman J.L.2000. "Silphium, Silver and Strife: A History of Kyrenaika and Its Coinage" *Celator* 14 (10): 6-24

Theodorides J. 1968. "Historique des connaissances scientifiques sur l'abeille" in Chauvin R. (ed). *Traité de biologie de l'abeille*. Masson. Paris, V, 2-34

Thomsen ML. 1992. "The evil eye in Mesopotamia" *JNES* 51(1):19-32

Topalidis N. 1968/1969. "Η μελισσοκομία στην Σάντα" *Αρχείον Πόντου* 29:332-339

Triopmhe R. 1989. *Le lion, la vierge et le miel*. Les Belles lettres. Paris

Tselios D. 1997. "Αξιοποίηση άγονων εκτάσεων με την ανάπτυξη της μελισσοκομίας" *Μελισσοκομική Επιθεώρηση* 11(6):268

Tsountas C. 1908. *Αι προϊστορικαί ακροπόλεις Διμινίου και Σέσκλου*. Βιβλιοθήκη της εν Αθήναις Αρχαιολογικής Εταιρείας. Athens

Typaldos-Xydias 1927. *Η νομαδική μελισσοκομία εν Ελλάδι*. Ελληνική Γεωργική Εταιρεία. Athens.

Tzavella-Evjen C. 1970. *Τα πτερωτά όντα της προϊστορικής εποχής του Αιγαίου*. Η εν Αθήναις Αρχαιολογική Εταιρεία. Athens

Tzedakis G., Martlow H. (eds). 1999. *Μινωιτών και Μυκηναίων γεύσεις*. Ministry of Culture and National Archaeological Museum. Kapon. Athens

Vandenabeele F., Olivier I.P. 1979. "Les idéogrammes archéologiques du linéaire B" *Etudes Cret* 24. Paris

Varella E. 2000. "Μελίκρατα και οξυμέλιτα στην ελληνική θεραπευτική" in *Η μέλισσα και τα προϊόντα της*. Πολιτιστικό Τεχνολογικό Ίδρυμα ΕΤΒΑ. Athens. 221-231

Vollgraff W. 1909. *Nikander und Ovid*. Groningen.

Vrontis A. 1939. "Η μελισσοκομία και το μαντρατόρεμα στη Ρόδο" *Λαογραφία* 12 (2), 195-208

Walberg G. 1986. *Tradition and innovation: Essays in Minoan Art*. Philipp von Zabern. Rhein.

Warren P. 1984. "Of Squills" in *Aux origines de l'Hellénisme. La Crête et la Grèce. Hommage a Henri Van Effenterre*, Paris. 17-24

Weingarten J. 1990. "Three upheavals in Minoan sealing administration: evidence for radical change" in Palaima G. (ed) Aegean Seals, Sealings and Administration *Aegaeum* 5, Universite de Liege, 105-114

Weniger "Melissa" in Rocher W.H. *Ausfuhrliches Lexicon d. griechischen u. romischen Mythologie* (1884 -), 2637-2643

West M.L. 1985. *The Hesiodic catalogue of women.* Clarendon Press. Oxford

Wheler G. 1682. *A journey into Greece by George Wheler Esq. in company of Dr. Spon of Lyons.* London

Wickens J. 1986. *The Archeology and history of cave use in Attica, Greece from prehistoric through late roman times.* Ph.D. Thesis. Indiana University. vol. I

Willetts R.F. 1962. *Cretan Cults and Festivals.* Routledge and Kegan Paul. London

Willetts R.F. 1977. *The civilization of ancient Crete.* Botsford. London

Willetts R.F. 1985. "Bee-keeping and Bee cults (Mallia and Ephesos)" Πεπραγμένα του Ε' Διεθνούς Κρητολογικού Συνεδρίου (Agios Nikolaos, 25.8-1.10 1981), Detorakis, Theocharis, (eds). Vol. A'. Herakleion, Crete, Εταιρεία Κρητικών Ιστορικών Μελετών. 407-412

Wingerath H. 1995. *Studien zur Darstellung des Menschen in der minoischen Kunst der alteren und jungeren Palastzeit.* Tectum Verlag. Marburg

Winston M.L. 1987. *The biology of the honeybee.* Harvard University Press. Cambridge, Massachusetts

Woudhuizen F.C. 1997. "The bee-sign (Evans No.86): An instance of the Egyptian Influence on Cretan Hieroglyphic" *Kadmos* 36:S.97-110

Zafiropulo J. 1966. *Mead and wine. A history of the Bronze Age in Greece.* trans. Green P. Schocken Books. New York

Zammit-Maempel G. 1968. "The evil eye and protective cattle horns in Malta." *Folklore* 79(1):1-16

Zymbragoudakis C. 1979. "The bee and beekeeping of Crete" *Apiacta* 14(3):134-8

www.ingramcontent.com/pod-product-compliance
Lightning Source LLC
Chambersburg PA
CBHW061544010526
44113CB00023B/2791